T0213523

New Paradigms in Healthcare

In the first two decades of this new millennium, the self-sufficiency of Evidence-Based Medicine (EBM) have begun to be questioned. The narrative version gradually assumed increasing importance as the need emerged to shift to more biologically, psychologically, socially, and existentially focused models. The terrible experience of the COVID pandemic truly revealed that EBM alone, while being a wonderful scientific philosophy and containing the physician's paternalistic approach, has its limitations: it often ignores both the patient's and physician's perspectives as persons, as human beings; it pays relentless attention to biological markers and not to the more personal, psychological, social, and anthropological ones, removing the emotions, thoughts, and desires of life, focusing on just the "measurable quality" of it.

Aim of this series is to collect "person-centered" contributions, as only a multidisciplinary and collaborative team can meet the challenge of combining the multiple aspects of human health, as well as the health of our planet, and of all the creatures that live on it, in a common effort to stop or reverse the enormous damage committed by humans during our anthropocentric era: a new paradigm of healthcare, education and learning to create a sustainable health system.

More information about this series at https://link.springer.com/bookseries/16769

Maria Giulia Marini · Jonathan McFarland
Editors

Health Humanities for Quality of Care in Times of COVID -19

Springer

Editors
Maria Giulia Marini
Fondazione ISTUD
Milano, Italy

Jonathan McFarland
Faculty of Medicine
Autonomous University of Madrid
Madrid, Spain

Sechenov University
Moscow, Russia

ISSN 2731-3247 ISSN 2731-3255 (electronic)
New Paradigms in Healthcare
ISBN 978-3-030-93361-6 ISBN 978-3-030-93359-3 (eBook)
https://doi.org/10.1007/978-3-030-93359-3

This Springer imprint is published by the registered company Springer Nature Switzerland AG
The registered company address is: Gewerbestrasse 11, 6330 Cham, Switzerland

Preface

Dear reader, how would you feel about going on whatever YouTube link is available free online to listen to Shostakovich's 7th Symphony, "Leningrad"? Would you like to try? Or if you have Spotify or even the chance to have a good classic edition in vinyl? We promise that this has nothing to do with your knowledge of Classical Russian music of the XX century but to use this music as a Language of care despite the obvious fact that the taste in music is extremely subjective.

Let's imagine that we have been able to persuade you to play the symphony up to the minute 7.33, after an impressive and dramatic opening with tiny inserts of the delicate flute, the music continues with its ups and downs. And then, at minute 7.33 "that March" begins. It is "that March" because it was meant to mimic the Nazi siege of the beautiful city of Leningrad (the Venice of the north) in 1941. The march blossoms amongst drums and oboes, and flutes, and it enchants the ear, leaving a "meme" in the brain that it will persist for hours. All the symphony is in C major; it is witty, triumphant, crazy, and overwhelming. At minute 23 the march stops and there is a nostalgic opening to what Leningrad was and is not anymore, however always with a rounded sonority of a major tune.

Dmitri Shostakovich's Symphony No. 7 in C major, Op. 60, nicknamed the Leningrad was begun in Leningrad, just before the Nazi siege, and completed in the city of Samara. The Leningrad soon became popular in both the Soviet Union and the West as a symbol of resistance to fascism and totalitarianism, thanks in part to the composer's microfilming of the score in Samara and its clandestine delivery, via Tehran and Cairo, to New York, where Arturo Toscanini led a broadcast performance and *Time* magazine placed Shostakovich on its cover. This work is still regarded as a major musical testament to the 27 million Soviet people who lost their lives in World War II, and it is often played at Leningrad Cemetery where half a million victims of the 900-day Siege of Leningrad are buried.

Beyond the facts, in Leningrad people were dying, starving, freezing; promised food supplies not arriving with endless desperate citizens boiling leather to eke out some nourishment, eating rats became a daily occurrence as well as burning chairs and beds to produce a flicker of heat.

In Leningrad, during the siege of the winter of 1941, the local Russian orchestra had one precise task through radio: to console, to ease the pain and suffering of the population and bringing ease and peace. That symphony was already being played around the world, but this particular orchestra of 18–20 men and women, who were starving to death, knew how to play the music, and broadcast it on the local radio to portray the feelings of living under siege. That symphony is not always easy to listen to, it encounters hurdles, discontinuities, and enigmatic sonorities; it screams, it cries, it asks for helps, it becomes cacophonic, but also after few seconds there is a wonderful harmonic piece and peace. It gives the idea of tremendous turmoil, of trying to escape, of searching desperately for food in this forced lock down that lasted for almost 3 years. On the other hand however it portrays friendship, resilience, beliefs, and ethics.

When all the authors wrote their chapters during this pandemic, that somehow metaphorically could be called a siege, without us being aware of it, we fell into a similar kind of epic genre that characterized the music of Shostakovich. All of us have been writing our chapter with a style ranging from epic to dramatic, most of us with an ending of dawn, of future, or as the musicians would say, with an ending in C major. There was no other possible genre, writing a book on Humanities for Health during the pandemic time to honour the people who died of COVID-19 (around four million deaths at the moment, but probably so much more), the healthcare professionals who died and their fatigue, and all the other people who died since they had no access to care. Beyond this huge loss, the sacrifices imposed on every one of us who had a very limited freedom, the impossibility of human touch, like in a dystopian novel in which we were catapulted, locked inside, with "the key thrown away". Drama and epic and quest are pervasive in all these chapters, some more technical in a less coloured way, and other like anthems to commemorate and celebrate the positive facets during these periods.

The acknowledgements have to go to our resilience, our strengthening, and our willingness to speak about what has happened (is happening) in a dedicated way, and to Springer who accepted publishing our reflections.

Parallel with music, please, go to the final minutes of the Symphony n° 7 (at 1 h and 11 min) and listen to the "crescendo", and imagine the immense scientific discoveries, the vaccines, the facts that for many countries the economic debts were cancelled, the reconciliation with nature, the efforts of all volunteers, and listen up to the end of the Leningrad symphony, which has a triumphal but not rhetoric final musical landscape.

In February 2022 the scenario which should have been the final chapter of the pandemic narrative, a renaissance of life and hope for all humanity, we now find ourselves once again embedded in another catastrophic global tragedy: war. This requires further resilience, cooperation, wisdom and courage. We, as co-editors, on behalf of all the contributors to this book that is dedicated to improving caring for patients, healthcare professionals, and all citizens wish to add the following lines by Mahatma Gandhi, "Non-violence is the greatest force at the disposal of mankind. It is mightier than the mightiest weapon of destruction devised by the ingenuity of man".

Acknowledgements in this pandemic with an uncertain ending go to the international network of health humanists, who were challenged by writing this book, to EUNAMES (European Narrative Medicine Society) to The Doctor as a Humanist. We also wish to acknowledge all the citizens who stayed at home and helped in this manner to diminish the contagion, to those who were working tirelessly providing food for us while we were in lock down, to the volunteers, to the old people who died and those who survived, to the teachers who continued distance learning and teaching the students, to the children who stopped playing in the park, to the scientists who are still trying to understand how to get a possible cure and better vaccines to cover the variants (now the Delta is the most terrible, but nobody knows what will happen in 2 months, 5 months, or when this book is published).

Here the names of the people we wish to thank, who contributed with their chapters and the glossary of this book, are as follows: David Cerdio, Paola Chesi, Carol-Ann Farkas, Fabrizio Gervasoni, Azar Ghasemi, Stephen Legari, Susana Magalhaes, Saba Mirikermanshahi, Negin Nouarei, Albina Vegel, Neil Vickers, Mahdi Azadi Badrbani.

It is very difficult to remember all those who have helped in the making of this book but here is an attempt to do just that since, even though in the words of Jonathan's paternal grandfather writing in a similar moment just after the second world war, "the times are troubled and trying: the present appears treacherous and the future uncertain" [1], there have been great moments of solidarity and companionship and of a spirit of pulling together. He himself continues in this vein of hope and spiritedness when he says, "But if we put our doubting mind in its place, and if instead we search the hearts within us, we shall find a sense of inevitable success and a feeling of ultimate triumph". So, with these life-affirming words, I, Jonathan McFarland, wish to mention a few of those who have supported us with this spirit. As well as the whole The Doctor as a Humanist team (and especially the members of the steering committee: David Cerdio, Irina Markovina, Joan B. Soriano, Joaquim Gea Guiral, Jacek Mostwin, Manuel Pera, Mariam Pogosyan, Theodora Tseligka, Ourania Varsou, Jose Marcos Felix Castro, Josep Eladi-Baños, and Alfredo Quiñones-Hinojosa), students, and supporters, I particularly wish to thank Professor Irina Markovina for her continuous support and friendship, and for giving me my first big break, and to Dr. Julio Ancochea and Dr. Joan B. Soriano for their unflagging belief in this and other projects since their enduring faith in me has been critical. I also wish to mention those who have read and helped with ideas for parts of the book, giving that essential "extra" pair of eyes, and they are Dr. Antoni Truyols, Gerard Bibiloni Isern, Josie Hughes, and Aina McFarland.

But I would not even have considered tackling this difficult project at this difficult time (or any for that matter) without the continued support and love of my wife, Polimnia, and daughters, Aina and Bianca, who are the pillars that sustain me and give me fortitude. And, of course, to my mother and father who taught me never to give up and imbued in me an inherent sense of enthusiasm, which makes or impels me to embrace what others might believe impossible, or at least foolhardy.

I, Maria Giulia Marini, wish to thank all the other friends of the European Narrative Medicine Society, EUNAMES, June Boyce-Tillman, Bernardino Fantini, Isabel Fernandez, Pilar Lorente, Sorin Costreie, John Launer, Angelika Messner, Weronica Madryas, Stefania Polvani, Antonio Virzì, Federica Vagnarelli, Ourania Varsaou, Mauro Zampolini, Weronica Madryas, and the others affiliated (from phylos, friends to come). A special thanks to Anna Wierzbicka and Cliff Goddard discoverers on the Natural Semantic Metalanguage. My deep thoughts and nostalgia go to Bert Peeters, who passed away in February 2021, with whom we have opened this new way, mixing linguistic to narrative medicine. My praise goes to my colleagues at the Wellbeing and Health Care Area of Fondazione ISTUD who continued to work ceaselessly to listen and to help with the research and education of the Healthcare situation throughout the whole pandemic time: Luigi Reale, Paola Chesi, Antonietta Cappuccio, Alessandra Fiorencis, Eleonora Volpato, Delia Duccoli. And to my family, fragmented into three different parts of the world because of this pandemic: Italy, Sicily (different from the continent), and UK. A new philosopher is born during the pandemic time; my son Gabriele with whom the dialogue now is not only affective but extremely witty and curious; my mother Giulia, incredibly resilient after an infarction on May 2020, who shielded herself, hoping that the new generations could catch up and move on, my sister Marina and her beautiful daughter Evelyn, in London, my sister Lucia who had a wonderful baby girl, Sofia, right during the pandemic and my "soul" sister Letizia, who lives in Sicily. To all the other relatives of my innovative family. To my friends, Cristina, Monica 1, Monica 2, Gabriella, Mario, Cinzia, Marco; together we were the force, the energy to move on in this turmoil. To my home, which sheltered us, and to the home gardening in the little balconies. Thanks to technology which helped us to remain in "touch". Zoom, Skype, teams, WhatsApp, Facebook….

Going back one century, I think of two other women; my great grandmothers who both died in their prime from the Spanish Flu. Here we are now, imagining giving voices to them as if they were alive. Differently from COVID-19, that flu was silenced for political reason: in our current global world, there are disadvantages such as the virus spreading, but also advantages, the fact that it cannot be hidden so quickly. Indeed, including our forefathers' and foremothers' roots, we should embrace the future generations, those who come after us, and aspire to leaving a better future.

Milano, Italy	Maria Giulia Marini
Madrid, Spain	Jonathan McFarland

Reference

1. McFarland BL. The will to live. Liverpool Philomathic Society; 1949.

Contents

Introduction

The COVID pandemic has led us into an upheaval that has made us question the certainties underlying what it means to be a human being in our age: the ability to control medical and social facts through evidence. For the first time, Western and developed countries have had to confront what many populations from the developing world (Africa, Latin America, etc.) face on a daily basis with HIV, Ebola, etc.

The Interconnectedness of Globalization has been the real disseminating catalyst of COVID-19, and many scientists wonder if this virus is the result of the technical Anthropocene age, with its indisputable lack of respect for the natural ecosystems. The virus has demonstrated that our frailty is only skin deep, and it has not only brought death and despair, but it has broken our interdependency as human beings, by imposing self-isolation as well as creating new ways of connections so that safety cannot imply loneliness.

In this book written between Winter and Spring of 2021, so almost 1 year after the explosion of the pandemic event, the coping strategies that may originate from the intersection of disciplines and sciences, such as narrative, literature, medicine, clinical science, natural sciences, philosophy, art, digital science, and linguistics, are used not only as reflective tools to promote health but also as a tension of well-being and wider awareness amongst carers, patients, students, and citizens of our planet, the Earth. These coping strategies should be supported by the decision-makers since they are the low-cost investments necessary to make the healthcare systems work; some of which have already been included into the political "resiliency plans" displayed by different governments, but with no particular educational policy on how to achieve them. We know how COVID-19 has revolutionized our lives: we had to look back to the past and learn from ancient wisdom of *impermanence* and at the same time boost the present with new scientific discoveries, and learn to savour art and literature as soothing balms.

All of this requires a change of cultural paradigms: to some extent COVID-19 acted as an amplifier revealing the intrinsic limits of our imperative reductionist biomedical model, applied to the healthcare services often deeply scarred by almost 20 years of spending reviews (budget cuts), with the remains of resources allocated mainly to non-communicable diseases, with no investment for infectious diseases, deemed short-sightedly as far away from our Western worlds. The healthcare system has to be profoundly reorganized as well as prioritizing the protection and empowerment of their professionals, who after more than 1 year have been seriously battered by the pandemic.

A good carer needs to heal themselves to be a good healer. The COVID experience has paradoxically created an unexpected but opportune moment for all carers since this unwanted trauma could be the fundamental ingredient to becoming a good carer. But this imbalance of certainties, fear, and loss of scientific power has to be elaborated and debated to reformulate the health service for humans and their ecosystem with new and renewed energies.

This is a book for all healthcare professionals, and the policy makers, which aims to create a New Paradigm of culture by intermingling the tiles of the SHAPE (Social Sciences, Humanities and Art, for People, Economics) disciplines with the STEM (Sciences—with its clinical side—Technology, Engineering, and Mathematics) ones. But it is also reading for patients and citizens of how to live better with and within this new world, which looks longer than originally thought: we seem to be living in a dystopian novel, with the ending unknown and in which we have to rebuild our personal, social, and professional identities, in our health and social care.

The authors unveil a polyhedric proposal of languages which can heal, soothe, and care for our minds, souls, and bodies. For the first time, in the COVID age, Science with its Evidence-Based Medicine is being analysed jointly with experts from the humanities of health in order to try and make some sense of the reaction to the pandemic event.

The international authors of the chapters of this book have been connected online since the beginning of the pandemic (from the end of February 2020) ceaselessly trying to provide some hope and logical solutions to the dilemmas and anguish of this dystopia by using tools for a new reawakening of the healthcare services, and wider still, the whole of our global society.

The book has been thought out as a journey, from the stunning and breathtaking explosions of the first outbreak of COVID-19 to the declaration of the pandemic by the WHO and the endless emergency activities organized to limit the catastrophic event, not only from the biomedical point of view but mainly from the psychological, social, and spiritual dimensions.

Voices of philosophers, epidemiologists, teachers of literature, scientists, scholars of narrative medicine, of planet sustainability, art therapists, physicians, and counsellors are included here as authors of the chapters: but present with them is the voice of the history of the last century, moving from the Spanish flu to our contemporary pandemic, in a mixed writing style which bridges quotations, comments, press releases, scientific publications, essays, and reflective thinking.

Almost every chapter has a practice time, which is not to be intended as a "*mandatory to do task*" but is more concerned with eliciting a spirit of curiosity, creativity, and inner knowledge. More than this, it would be great if it could be considered as a game since play always causes the release of endorphins, the natural painkiller which can lower stress and anxiety levels, and even creates a sense of euphoria. This combined with other neurotransmitters may help create an ideal environment for focused learning.

Fig. 1 Photo by Antoni
Truyols

   As an appendix there is a structured glossary which describes the words of the world of Humanities for Health and of the Pandemic.

   And then, at the end of the book, there are few empty pages so that every reader can write the thoughts, emotions, projects, and wishes that any particular chapter may conjure up in a free association manner. We hope that by reading this book we may help you, the reader, to learn to live and perhaps prosper with the unknown, the mysterious, moving from the microcosm of the virus to embrace the macrocosm of the universe, looking outside the window (Fig. 1).

Maria Giulia Marini
Medical School, La Sapienza, Roma, Italy
Fondazione ISTUD, Milano, Italy

Jonathan McFarland
Faculty of Medicine, Autonomous University of Madrid, Madrid, Spain
Sechenov University, Moscow, Russia

The Grand Narrative of COVID-19 Pandemic, Among Health, Science, History, and Citizenship

Maria Giulia Marini

> Modern science has been a voyage into the unknown, with a lesson in humility waiting at every stop. Many passengers would rather have stayed home.
>
> Carl Sagan

After the outbreak of COVID-19 in Wuhan, in China, the World Health Organization (WHO) between January 10 and 12 published a comprehensive package of guidance documents for all countries, covering topics related to the management of an outbreak of a new disease: this new, unknown, and difficult to classify disease landed in Italy ("officially" first Western country) on February 20th 2020, and, a few weeks later it was spreading all over the country, becoming a pandemic, as declared by the World Health Organization on March 11th (this number, somehow by now, a symbol of bad omen days) 2020:

> In the past two weeks, the number of cases of COVID-19 outside China has increased 13-fold, and the number of affected countries has tripled. There are now more than 118,000 cases in 114 countries, and 4,291 people have lost their lives. Thousands more are fighting for their lives in hospitals. In the days and weeks ahead, we expect to see the number of cases, the number of deaths, and the number of affected countries climb even higher.
>
> WHO has been assessing this outbreak around the clock and we are deeply concerned both by the alarming levels of spread and severity, and by the alarming levels of inaction. We have therefore made the assessment that COVID-19 can be characterized as a pandemic.
>
> Pandemic is not a word to use lightly or carelessly. It is a word that, if misused, can cause unreasonable fear, or unjustified acceptance that the fight is over, leading to unnecessary suffering and death. Describing the situation as a pandemic does not change WHO's assessment of the threat posed by this virus. It doesn't change what WHO is doing, and it doesn't change what countries should do.

M. G. Marini (✉)
Medical School, La Sapienza, Roma, Italy

Fondazione ISTUD, Milano, Italy
e-mail: MMarini@istud.it

> We have never before seen a pandemic sparked by a coronavirus. This is the first pandemic caused by a coronavirus [1].

These were the words used by the WHO leaders, which came out with a perceived slight delay when COVID-19 was already a reality in many countries. This delay in communication was possibly due to the danger of "escalation", moving from the word "outbreak" to "pandemic", that comes from Greek, *pan* (the whole) and *demos* (people); the entire population of the world. On that very day, the stock markets had their most dramatic fall of the last decade, which is a clear demonstration of how finance and economy are linked to countries' health and welfare.

After a few weeks, it appeared that, for most inhabitants of this planet, that Normality had returned before the end of the year and that a New Normal had to be pursued and found: the divide was immense, between an *Ante—Covid age*, with an *Ante Covid* life, and an *In—Covid age*, with an *In -Covid* life, although many of us hoped that we would be able to close this matter as soon as possible, projecting us into a *Post—Covid age*.

While writing these lines in March 2021, those who have been vaccinated in Europe range from 1.4% to 7.4% of the population, varying among countries. The situation is far from being under control, with the daily variants of the virus popping up from different laboratories located around the world [2]. Thus, the *In Covid* age is truly open.

In this chapter, we will explore the grand narratives of COVID-19, the metanarratives of different professional populations, and of the citizenship narratives, with scientific, historic, and linguistic references. Metanarratives [3] is a mandatory method since we are dealing with a pandemic event that may impact 7.8 billion people; there will be many limits to this grand narrative, losing the originality and uniqueness of so many individuals' narratives which could be locally collected.

However, some "universal" or- "pandemic"- words are tracked and emerged from local narratives, during the SARS-COV2 world infection.

From the Spanish Flu to the Covid Age

> All mankind is one author, and is one volume, when one man dies, one chapter is not torn out of the book, but translated into a better language, and every chapter must be translated: God employs several translators; some pieces are translated by age, some by sickness, some by war, some by justice. But God's hand in every translation and his hand shall bind up all our scattered leaves again, for that library where every book shall lie open to one another.
> John Donne, Meditation, XVII

In the first announcement of the SARS- Cov2 pandemic by WHO, a word is very evident: *fight*. It is the same word used to declare how to combat cancer and many other diseases. However, contrary to a chronic illness impacting on only one individual, for the first time since the Spanish flu of 1918–1919 that only lasted two years, the whole world and all its human beings is at risk "of fighting" this disease. It is impossible to compare the acceleration of the time in the third millennium, in the few years ante COVID-19 age, with our attitude to "devour time", to plan works and holidays, to live globally flying everywhere in a liquid society characterized by rampant consumerism, [4] with the times of the Spanish flu. At that time, indeed, there were massive migratory movements, but much slower than now; war, more poverty, possible undernourishment were the possible causes of the debilitation of the young people who died of this "forgotten pandemic" [5].

Despite the death count due to the Spanish flu accounting for an uncertain number from 40,000,000 to 100,000,000 million people in the world (4% of the population), the power of that "historic" virus was obscured by other headlines: World War I and its tragic sequelae. Amongst soldiers and civilians that war killed approximately 40,000,000 people in parallel to the pandemic. Memorials built in the winning countries celebrated the soldiers of this war and very few monuments were erected to trace memory of the immense power of the deadliest flu. However, when we were children, in our constant family narratives, we heard directly from our parents and grandparents about the Spanish flu. It was tacit hidden knowledge, carried within us in our ancestral tree.

Many epidemics have occurred during the last 50 years, probably more frequently than in the centuries before, and above all we should mention the 2002–2004 SARS outbreak caused by severe acute respiratory syndrome coronavirus (SARS-CoV or SARS-CoV-1). The outbreak was first identified in Foshan, Guangdong, China, on 16th November 2002. Over 8000 people from 29 different countries and territories were infected, and at least 774 died worldwide. The major part of the outbreak lasted about 8 months until the WHO declared that it had been contained on 5 July 2003. However, several SARS cases were still being reported until May 2004. According to the UN Department of Global Communications, the Asia-Pacific region was especially prepared against COVID-19 because of its experience and knowledge of handling outbreaks. And, as stated in a WHO report of 2020, the region has made significant progress in health security systems in the past decade. These systems were built from lessons learned from the SARS epidemic, the H1N1 influenza epidemic, and real-life events, leading to the development and implementation of the 48-nation Asia Pacific strategy against emerging diseases and public health emergencies.

In Western countries, however, the daily clinical practice paid less attention to infectious diseases, and this is a possible reason why we did not fear the potential "next big one", the viral infection, as the virologists and epidemiologists were calling it [6] and warning the health care system to be prepared to face a massive pandemic event. We were glancing at risks of possible or real wars, with human enemies, in a sort of cognitive distortion, not listening to voices of the scientists who were considered to be talking from "outer space". Odd, since in the history of mankind, as written above, not only wars but also pandemics due to viruses and other infectious agents have continuously afflicted humanity. With this shortsighted approach and distinct lack of humility towards other guests of the planet, the viruses, human beings in the institutional scientific bureaus considered the risk of pandemics in the western world as very remote; "exiling" them to developing countries. AIDS somehow successfully kept under control with anti-AIDS drugs, and SARS extinguished. The seasonal flu had its rhythm of vaccines. Genetics, with its genomic and precision medicine, was mainly focused on forecasting if the cause of death would be a cardiocerebrovascular event or cancer [7]. No other causes of death, in our welfare state, apart from sudden death like war, or suicides, or external accidents, were possible: but, alas, a foreseen pandemic event came to change our normal life. I must stress the word "foreseen" since the infection of COVID-19 was forecast although it was impossible to know what kind of disease would appear.

Globalization is indeed a huge factor for the accelerating and spreading of the virus; since the human population has reached 7.8 billion [8], and as an increasing number of us travel with greater frequency across the whole world, from continent to continent, the potential reach of infectious disease outbreaks has been vastly magnified. The chances for a new virus to spill over from an animal like a monkey or bat into other animals, humans, has also increased as humans have intruded into the natural habitats where these creatures live. When rainforests get cut down, displaced wildlife is forced to find food and new homes closer to villages and towns, multiplying the opportunities for exposure. David Quammen, award-winning science, nature, and travel writer, in his masterpiece essay "Spillover" published in 2012, forecast the pandemic. His words are: *"When I first read about Ebola, the notion of getting on a plane and going to the place where that virus lived seemed to me foolhardy to the point of suicidal. The more I learned about these diseases, the more my irrational fears were replaced by rational concerns. With every new virus, biologists and infection control doctors learn more. Science has advanced almost as quickly as human travel patterns have expanded. Forecasting and diagnosis of new viruses have greatly improved, incorporating new technologies."* Ultimately, what Quammen wants to leave readers with is not an alarmist fear of animals and disease, but a more comprehensive portrait of an interconnected world. *"One of the salubrious things about infectious disease is that it reminds people that we are not separate from nature,"* he says. *"It's a reminder of our connectedness to the rest of the living creatures on planet Earth."*

Only a part of science could have been prepared, but not the vast majority, and on top of this, the vast majority of the health care services were suddenly overwhelmed by the massive force of this pandemic. However, what we should take with us of the studies and research made by Quammen is that we should stop from using the words "fight"- combat more related to body contact- or "war"- a contention which implies a military hyperarchy and technological weapons. We are interconnected with the cosmos, and therefore, the belief that will help us relies on "interdependence", not only with other human beings but also with the planet itself. Viruses have driven close to 30% of all adaptive changes for humans and they are one of the most dominant drivers of evolutionary change across mammalian and humans [9]. The appearance of Homo Sapiens caused however a huge reduction in the biodiversity of animals and plants, and with it, increasing the risk of illness from zoonosis.

If we accept this grand narrative, that encompasses our planetary belonging, we can start a more responsible journey into this new "landscape", "mindscape", and "timescape". Fight and war are metaphors for survival: however, the "live translators"- as John Donne, the English poet, calls the professions used by God, taking away any transcendental push, could be science with our brain and heart, the cycle of our ecosystem and the pattern of history. They may hold surprises and discoveries; responsibility and interconnection, as well as inner and outer journeys.

John Donne wrote his meditation on death contextually to his historical period: the death penalty was there, people were dying for justice, and war; interestingly, the poet kept separate sickness from age, and this opened the discussion to which limit aging can expand without death.

It is well known that COVID-19 hits the most vulnerable and fragile people, especially the elderly. When John Donne was alive, viruses had not been discovered, the lifespan was much shorter, and it was normal to die of aging. "Long live the king, long live the queen"; these were the blessings, not the western and infantile "request for immortality" of the liquid society. In 2020, in our post-modern society, the western countries were not prepared to face the virus, not prepared to see so much death in elderly people: it is terribly true that in some cities and towns, the eldest generation was almost wiped out, with no possibility of mourning those that had passed away with a funeral ritual, with consequences for the passage of the legacy to the left-over generations. The deconstruction of death through medical intervention has been a pillar of the modern age, towards the utopia of the immortal life with the biomedical model being a catalyser of this, but Covid-19 disrupted this abruptly, especially that of longevity which could be stretched to eternal life.

After one year *In Covid Age*, short narratives, fast impulsive thoughts have left time to the Grand Experience Narrative of rethinking the priorities of our lives, and our relationship with others and with the planet: new meanings are there to bring solace and comfort to the death, limitations, and sacrifices that we are all experiencing. In the words of the WHO on March 8th, *"We have never before seen a pandemic sparked by a coronavirus"*, there is an emphasis on the new condition and the unseen, therefore the unknown, the mystery.

In this responsible journey, the never before seen tremendous scientific must be highlighted as a terrifically positive and mean-making.

The Grand Narrative of Evidence-Based and Narrative Medicine

The interplay between real evidence-based data and the people's narratives was fundamental. We can never only be reducible to numbers of infected people, swamps, hospitalized patients, pneumonia, disseminated intravascular coagulation, cortisones, hydroxychloroquine, low weight heparin, oxygen, and hopefully vaccines - all fundamental markers and discoveries to take into consideration in the narrative of Covid-19, since so less is known about this virus. The human voice echoes the facts that occurred in intensive therapies, in general practice, in the expressions of their emotions, thoughts, beliefs, and projects.

However, in this grand narrative of Covid-19, something "terrible" happened to Science: science has lost its capital letter, before the "unknown". The symbol of certainty in medical protocols fell to earth, embracing the unwished thinking that many doctors and scientists stated simply that they "did not know what to do". The first Covid studies were not evidence-based, full of biases, just the attempts by explorers on an unknown hostile planet.

Here are some of the open questions in such a prestigious medical journal as Circulation, in April 2020: "*While the pandemic is evolving its course, there will be a growing population of recovered patients. The majority will do well. However, many unanswered questions remain. Will exposed patients have adequate long-term immunity? Does the immunoglobulin G produced have adequate neutralization capacity? Those with complications may have a more challenging course of recovery and long-term sequelae. The immune activation and dysfunction can lead to target tissue fibrosis and microangiopathy, as was observed in some patients after SARS. This can affect long-term lung function, or if the heart is involved, residual cardiomyopathy…From an epidemiological point of view, an important question is how many people in the population ultimately acquire immunity to SARS-CoV-2 as antibody testing becomes more available. Will there be an effective vaccine for the remaining unexposed population? Will there be continued mutation of the virus? Does the virus have a natural reservoir? Are there enough asymptomatic carriers that can restart another infectious cycle?*" [10]. It was time to publish open questions to the scientific community and not recommendations: we do not yet have the answers to many of these open questions; however, we have developed vaccines. At that time, science was groping in the dark, with few clinicians announcing any certainties and many more scientists and practitioners shocked to see that current technology was not as effective as resuscitating therapy. It was, and still is, therefore, and will always be the time of grand narrative of questions.

There was no time to wait for long term results from clinical trials, although the first rule was, and still is, "*primum non nocere*", first do no harm. The grand narrative of the peer-reviewed journals as the Journal of American Medical Association, British Medical Journal, New England Journal of Medicine changed completely, publishing patients and case reports, even anonymous letters to denounce things not being done to prevent the pandemic situation, and trials with very few patients in *all in haste, and all for free*. The rhythm of the debate on the journals was hectic on the hypothesis that masks could prevent the infection when now no-one disputes this at all, what were the potential active drugs and which the maneuvers to perform, and the race for vaccines to test effectiveness and safety: hectic but always respectful.

In 2020, scientists were shattering their limits, creating a world of collaboration unlike anything seen in history. As never before, researchers had so many experts in so many countries focused simultaneously on a single topic and with such urgency. Normal imperatives like academic credit were set aside. Online repositories made studies available without waiting for the last approval of the publications. Researchers identified hundreds of viral genome sequences. An immense number of clinical trials were launched, bringing together hospitals and laboratories from around the globe.

As mentioned, "Science" lost its capital letter becoming "science" on the playground of the mediatic debate: scientists, virologists, epidemiologists, clinicians assaulted by journalists out of the hospitals. Up against the unknown ending of this pandemic, some started to dispute publicly, and they were pushed to release prophecies, somehow trapped into thinking they had a crystal ball to foresee the future: "it will be as the Spanish flu, with the second wave worse than the first", or "the virus will disappear at the end of the summer". This kind of anthropological grand narrative strengthened uncertainties amongst the medical class and decision-makers but did far more harm to the general public.

Anyhow, the evidence-based world shrank, and narrative medicine was very helpful in generating a hypothesis of contagion, of the link between air pollution-lung damage and Covid-19, on the hypothesis of possible drugs and on the use of vaccines. At the beginning of the outbreak, in shortage swamps, and still after swamps, to track the people who were met, the narrative was the only old method to know who was met, where, and when, even more than the dedicated app.

The bioethicist Alessandro Blasimme writes [11]: "*Many governments have seen digital health technologies as a promising tool to address coronavirus disease 2019 (COVID-19), particularly digital contact tracing (DCT) apps such as Bluetooth-based exposure notification apps that trace proximity to other devices and GPS-based apps that collect geolocation data. But deploying these systems is fraught with challenges, and most national DCT apps have not yet had the expected rate of uptake. This can be attributed to a number of uncertainties regarding general awareness of DCT apps, privacy risks, and the actual effectiveness of DCT, as well as public attitudes toward a potentially pervasive form of digital surveillance. DCT thus appears to face a typical social control dilemma*".

Beyond narratives using confidentially for tracing and tracking, individual narrative medicine was and is an extraordinary form of comfort and relief to unfreeze pain, sorrow, fear, and anger of the pandemic: many websites were and are collecting narratives of the facts that happened in the "free" democracies of the world. In some others, where citizens are not allowed to speak, narratives are shouted on social media, to be banned after a while.

Attention to language should have been mandatory in organizing educational campaigns on hygienic rules, translated into languages all over the world since we live in a multiethnic society, visual with simple written texts on the basic rules to follow would have reduced the risk of contagion. Nevertheless, these were only released (at least in Italy) at the end of April 2020 with a delay of almost two months. Ethno-linguists are also warranted to make sure not only to translate but to culturally validate the meaning of the rules across the populations as basic, and we could add, across generations: why are the highest "no mask" rates in young people despite the evidence shown that they are spreaders of the virus? Narrative medicine could help in understanding the reason for this non-acceptance of the rules and to design more effective campaigns, tailored for the young generations.

The Citizenship Grand Narrative

From an independent study of Fondazione ISTUD, more than one hundred narratives in Italy from students, adult, and elderly people locked in their houses were collected from March to the end of April 2020. Proceeding from the low ranking to the number one used metaphor, the fourth group of the most recurring metaphors was that of images that described the sense of perceived "destruction"; the arrival of the virus was seen by many as a natural sudden and disruptive catastrophe, like a tsunami, a tycoon: "We have all been overwhelmed by a tsunami", "the pandemic broke out which affected all the services hospital", "certainties can be swept away in no time". More, there were frequent images of the ugly "nightmare" from which we hope to wake up: "It seems to me to have been catapulted into a nightmare", "I would like to wake up in the morning and get out of this nightmare"; feeling "inside an apocalyptic film", "the hysteria that storms supermarkets like in a post-apocalyptic film", "inside the movie Lethal Virus"; the "wound" that comes to be perceived in the bodies: "I felt a blade that entered my side", "someone even manages to smile at you the broken soul". Finally, the feeling of living inside a "bubble", "a parallel universe": "it seems that everyone lives his life, in a soap bubble", "as if on February 21, 2020, in Italy a kind of door space-time and had sucked us all into a kind of parallel universe". The dystopian world is narrated, where **Dystopia** (*dys*, bad and *topos*, place), is the representation of an imaginary reality of the future, predictable on the basis of highly negative trends of the present, in which an undesirable or frightening life is foreshadowed. Society is characterised by oppressive social or political phenomena, in conjunction with or as a consequence of dangerous environmental or technological conditions, including post atomic world wars, meteorites, glaciation, pandemics. The grand rising questions basically follow the research of the possible true: on one hand what the majority of science endorsed, that it was a spill over, with one potential place at the Wuhan Market. On the other hand, there was a minority of scientists and different politicians who were supporting the hypothesis of the laboratory leakage of a possible natural, or even worse, artificial virus: the report of the WHO [12] keeps on repeating that all the hypotheses are open. These are the words of Dr. Tedros, General Director of WHO, on March 30th, 2021, "*As far as WHO is concerned, all hypotheses remain on the table. ... We have not yet found the source of the virus, and we must continue to follow the science and leave no stone unturned as we do, finding the origin of a virus takes time and we owe it to the world to find the source so we can collectively take steps to reduce the risk of this happening again. No single research trip can provide all the answers.*" Hence, we should not be surprised at this point to read this sensation and have the feeling of living in a dystopian environment.

The third most recurrent image in describing the experiences of these first months of the pandemic was that of "imprisonment" in homes, for some considered as protection but for others called "cloister", "prison", "cage": "In the forced cloister I am confined to", "shopping has become a bit like the hour of air", "we are all at home under house arrest." This was and still is particularly remarkable in the younger generations but also for elderly people, in which there was an upgrading of the word "isolation" to "shielding", evocating protection, however with a shield, an object used during war.

The second group of the most represented metaphors was that of "time suspension", "the frozen world"—now a familiar concept to us, that we currently use during our on-line meeting when there is a line crash -; people's lives were suspended, "in a limbo", waiting to rediscover certainties about the future: "everything has frozen, inside and outside of me", "I live in a suspended time", "the whole world has paralyzed". For the first time in this new millennium, Time has acquired a different flavour in the citizens' grand narrative: due to the individual limits of movement, caused by the global lock down, the continuity of a routine such as going to school, or to the office, restaurants, gyms, cinemas, was stopped. The house (when available) and its perimeter became the only allowed space to live and spend time, mainly cleaning it, overwhelmed by the fear of contagion; journeys were not allowed, still more constrained. For health care professionals, the new routine was characterized by commuting between home and hospitals: all medical congresses with related journeys were cancelled and transferred to virtual journeys. Time has acquired a different meaning, having to do with a "bubble, lingering" effect. Many narratives reported the shock of moving from the hectic life of physical connections, made by flights, trains, and car transportation, to being imprisoned within a bubble. Although in many fields the business continuity was somehow guaranteed by the digital tools, time was mainly perceived as "frozen" since, apart from the ambulances, the devoted health care professionals, and the food and goods chain.

We all know quite well that the most used metaphor during the COIVD-19 pandemic was "war", mirroring the media and the health care professional language, therefore the realm of conflict.

Now, after one year of enduring COVID-19 and, although the vaccine campaign has started, we are living with the difficulties of an efficient logistic supply chain of vaccines, with daily announcements of new lock downs in many western countries. The policymakers and media narratives are still on the *war* side, every day broadcasting new infection rates and numbers of "victims", using the same term as if we were in a *war*. How long can we bear a war without being traumatized? Elena Semino is an expert in metaphor analysis [13], and she works on reframing the used metaphors to create healthier and more efficient images, and therefore beliefs, for the citizens, policymakers, and health providers. She proposes to switch from being soldiers to the metaphor of fire-fighters. Fire, she explains, conveys danger and urgency, distinguishing between different phases of the pandemic, and it explains how contagion happens and the role of individuals within. It also explains the measures for reducing contagion; portrays the role of health workers; connects the pandemic with health inequalities and other problems; and outline post-pandemic futures. In particular: *"Think of COVID-19 as a fire burning in a forest. All of us are trees. The R0 is the wind speed. The higher it is, the faster the fire tears through the*

forest. But just like a forest fire, COVID-19 needs fuel to keep going. We're the fuel."
Beyond the image of the fire, we can see here the image of being an interconnected
forest that can live in peace, if no harm has been done to our ecosystem, and there
are no major turbulent factors: we are an *entangled* community of livings.

Fire Fighters reminds us also the heroes of 9/11, who were photographed going
up the stairs facing certain death, when the flow of people was going down the stairs
in the World Trade Centre towers in New York. And this instantly recalls a terrorist
attack, with an external enemy able to cause the death of 2996 innocent people up
to 343 of whom, were, in fact, firefighters.

Learning from the Myth

Can we proceed further? Are there more powerful and effective models which can
help in this time and this place? I endorse the use of the myth, coming from the long
journey of the Odyssey. There is a man, Odysseus, who travels for twenty years in
search of his lost homeland. He has to face enduring trials, and, in the meantime, he
also learns how to live and learn from the sudden daily discoveries; watching the
future, he keeps a constant eye on the present, not in its "seize the day" format, but
in its wise, vigilant, strategic, emotional, affective, creative, and respectful attitude.
This man is Odysseus that Homer, the Greek poet, if he ever existed, defines as a
man from the "multifaceted genius": in Greek, the world "multifaceted" is literally
"multi-directional", "*polytropos*".

I think that we should go back to this immense mythological character, Odysseus,
who, during this long journey, is able to disappear, mask, reveal himself only when
appropriate. He makes many mistakes, indeed, but he learns from his errors.
Odysseus enters into the Polyphemus cave, the cyclops, desecrating therefore the
living of a wildlife creature. This is his biggest error for two reasons: the first con-
tingent reason is that Polyphemus eats the human flesh of his crew, and it is very
difficult to escape from the cave; the second strategic reason is that, after blinding
the cyclops, Odysseus will have to go against the power of Poseidon, the God of the
Sea and Father of the Cyclops. Thus, the journey of his return to Ithaki will become
longer for this mariner king.

Haven't we in the last centuries desecrated the life of our earth? Is there a possible causal relationship between what humankind has done to the planet throughout the centuries, and the contemporary increase in spillovers up to the generation
of COVID-19? Quammen says yes, David Attenborough says yes, [14], and I too
say yes; or very likely.

Here is the time, In Covid-19 ages, for a contemporary rethinking of our responsibilities during this life journey, our Odyssey, using multifaced genius, to protect
humankind and our planet, which is symbolized by the Whole Island of Ithaki. The
journey could be a long voyage with storms and high waves, but also with a calm
and serene sea, and dangerous shallow water again; by analogy, it could display
wonderful gifts, nice strangers could be met surfing the net, as well as scientific

Practice Time

1. In the frame "A pledge for planetary health to unite health professionals in the Anthropocene" published on the Lancet in November 2020 ([15], reproduced with permission) is presented:

   I solemnly pledge to dedicate my life to the service of humanity, and to the protection of natural systems on which human health depends.

   The health of people, their communities, and the planet will be my first consideration and I will maintain the utmost respect for human life, as well as reverence for the diversity of life on Earth. I will practise my profession with conscience and dignity and in accordance with good practice, taking into account planetary health values and principles.

   To do no harm, I will respect the autonomy and dignity of all persons in adopting an approach to maintaining and creating health which focuses on prevention of harm to people and planet.

   I will respect and honour the trust that is placed in me and leverage this trust to promote knowledge, values, and behaviours that support the health of humans and the planet. I will actively strive to understand the impact that direct, unconscious, and structural bias may have on my patients, communities, and the planet, and for cultural self-awareness in my duty to serve.

   I will advocate for equity and justice by actively addressing environmental, social, and structural determinants of health while protecting the natural systems that underpin a viable planet for future generations.I will acknowledge and respect diverse sources of knowledge and knowing regarding individual, community, and planetary health such as from Indigenous traditional knowledge systems while challenging attempts at spreading disinformation that can undermine planetary health.

   I will share and expand my knowledge for the benefit of society and the planet; I will also actively promote transdisciplinary, inclusive action to achieve individual, community, and planetary health.

   I will attend to my own health, wellbeing, and abilities in order to provide care and serve the community to the highest standards.

   I will strive to be a role model for my patients and society by embodying planetary health principles in my own life, acknowledging that this requires maintaining the vitality of our common home.

   I will not use my knowledge to violate human rights and civil liberties, even under threat; recognising that the human right to health necessitates maintaining planetary health.

   I make these promises solemnly, freely, and upon my honour. By taking this pledge, I am committing to a vision of personal, community, and planetary health that will enable the diversity of life on our planet to thrive now and in the future.

   What do you feel reading these words?

> *What do you think reading these words?*
> *What about the time in which this pledge was published, in the Covid Age?*
> *What about the respects of others Indigenous traditional knowledge?*
> *What about the taking care of the own health and wellbeing?*
> *I want that these words...*
> 2. *Would you like to write your individual Narrative after Covid-19 entered in the world and then in your country in 2020, up to now?*

discoveries, new therapies, cooperative networks, and beautiful ethics values represented by a spiritual inner and outer traveling.

References

1. Tedros Adhanom. Director of WHO. https://www.who.int/director-general/speeches/detail/who-director-general-s-opening-remarks-at-the-media-briefing-on-covid-19%2D%2D-11-march-2020.
2. https://www.euronews.com/2021/03/01/covid-19-vaccinations-in-europe-which-countries-are-leading-the-way.
3. Liotard J-F The post-modern condition. A report on knowledge. Minneapolis: University of Minnesota Press; 1984 [1979].
4. Palese E. Zygmunt Bauman. Individual and society in the liquid modernity. Springerplus. 2013;2(1):191.
5. https://wellcomecollection.org/articles/W7TfGRAAAP5F0eKS?gclid=Cj0KCQiA4feBBhC9ARIsABp_nbVXIs9l3QQcjzasveG-pTLsGOcw316y49jj34vb2QTjNuQ5j9Px4CgaAu1nEALw_wcB.
6. Quammen D. Spillover – animal infections and the next human pandemic. New York Time Press; 2012.
7. Aronson SJ, Rehm HL. Building the foundation for genomics in precision medicine. Nature. 2015;526(7573):336–42.
8. https://www.ft.com/content/e3ebd06a-0db8-11e2-97a1-00144feabdc0.
9. Enard D, Cai L, Gwennap C, Petrov DA. Viruses are a dominant driver of protein adaptation in mammals. elife. 2016;5:e12469.
10. Liu PP, Blet A, Smyth D, Li H. The science underlying COVID-19, implications for the cardiovascular system. Circulation. 2020;142(1):68–78.
11. Blasimme A, Vayena E. What's next for COVID-19 apps? Governance and oversight. Science. 2020;370(6518):760–2.
12. https://www.who.int/news/item/30-03-2021-who-calls-for-further-studies-data-on-origin-of-sars-cov-2-virus-reiterates-that-all-hypotheses-remain-open.
13. Semino E. "Not soldiers but fire-fighters" – metaphors and Covid-19. Health Commun. 2020;36(1):50–8. https://www.tandfonline.com/doi/full/10.1080/10410236.2020.1844989.
14. Attenborough D. A life on our planet: my witness statement and a vision for the future. Ebyry; 2020.
15. Wabnitz K-J, Gabrysch S, Guinto R, Haines A, Herrmann M, Howard C, et al. A pledge for planetary health to unite health professionals in the Anthropocene. Lancet. 2020;396(10261):1471–3.

Long Covid, Medical Research and the Life-World: A View from Bioanthropology

Neil Vickers

Introduction

This chapter asks if Covid-19 might change the way illness is perceived in the rich world. We live in a culture is that allows biology to overshadow every other way of relating to illness. But if medical history has taught us anything about pandemics, it is that they have cultural as well as biological after-effects. The fact that this pandemic unfolded in the internet age meant it could be studied differently from previous ones. Citizen scientists were mobilised in unprecedented numbers and in unprecedented ways to map the natural history of the disease. Ordinary sufferers played a key part in characterising the symptoms of Covid-19 in ways biomedicine had not anticipated, highlighting, for instance, the loss of taste and smell that affects some people with the virus, before researchers had had the opportunity to understand why this was so in the laboratory. And in a rare victory for patient-ethnographers, the phenomenon now known as 'Long Covid' was first described by people affected by the syndrome, rather than by professional Covid researchers. Drawing on a range of theories relevant to narrative medicine—medical and bioanthropology, social theory and history—I argue that Covid-19 and especially Long Covid offer a powerful opportunity to rethink the very nature of illness as a social and cultural phenomenon. It is, I suggest, an opportunity that must not be squandered. Taking as my starting point Elliot G. Mishler's notion those medical consultations are a site of dialectical struggle between 'the voice of biomedicine' and 'the voice of the life-world,' I pay particular attention to the possibility that Long Covid

N. Vickers (✉)
Department of English Literature, King's College, London, UK
e-mail: neil.vickers@kcl.ac.uk

© The Author(s), under exclusive license to Springer Nature
Switzerland AG 2022
M. G. Marini, J. McFarland (eds.), *Health Humanities for Quality of Care in Times of COVID -19*, New Paradigms in Healthcare,
https://doi.org/10.1007/978-3-030-93359-3_2

will serve to relegitimise a purely biomedical understanding of illness in general by ignoring or minimising the importance of the life-world. This would result in many sufferers from Long Covid falling victim to the stigma that is openly visited on people with medically unexplained conditions such as ME/CFS and covertly on others with serious physical illnesses. The chapter concludes with some proposals to strengthen the voice of the lifeworld.

Part I

Something very unusual happened in the second half of 2019. A new pathogen, severe acute respiratory syndrome coronavirus 2 (SARS-CoV-2) crossed the species barrier in Wuhan, and was transported in its human form to Western Europe and from there to North and South America. In all of these locations, the virus flourished. It reproduced itself on a scale unseen since the Spanish Flu virus of 1918–1919. Scientists sometimes talk about the emergence of new pathogens as instances of so-called 'ecological release'. Perhaps the most talked-about example in Western history is the so-called 'Columbian exchange' that occurred in the years following Columbus's arrival in the Americas when plants, animals, people, technology, diseases, crossed boundaries between Europe, the Americas and Africa [1]. As Alfred W. Crosby and others have shown, the arrival of the first colonists led to an ecological convulsion of a kind seldom seen on our planet. It has been compared with the extinction of the dinosaurs [2]. Columbus and his successors brought with them wheat, cows, goats, horses, sheep, dogs and cats. They transported a variety of European birds including ducks across the ocean. They also brought rats. Before their arrival, it is believed that there were no domestic animals in North America. (In South America, there were llamas and alpacas.) Consequently, the native population of the New World had very limited experience of, and were much more vulnerable to, zoonotic diseases, diseases that cross species barriers. The contrast with Europeans could not have been greater. The Europeans brought whole microcultures too, notably the insects and parasites that thrived invisibly on the surfaces of the exotic plants and animal bodies they brought for farming in the New World. And lastly, they brought diseases to which the indigenous human population had no preexisting immunity. These included smallpox and malaria. The traffic was not all one-way: syphilis almost certainly existed in North America before it was reported in Europe. But the death rate was massively skewed in the Europeans' favour. It has been estimated that in the 150 years following the 'Columbian exchange', somewhere between 80 and 95 per cent of the indigenous human population of North and South America was wiped out [3]. In many places, the entire population was exterminated. Crosby remarked that if Genghis Khan had invaded Europe around the time of the Bubonic Plague, the European population might have suffered a similar fate to the native Americans of the sixteenth and seventeenth centuries. Christianity might have died out and an Asiatic phase in European civilisation would have begun [1]. Some scholars have even suggested that the history of slavery in the New World was powerfully influenced by the Columbian exchange [4]. There was no malaria in the Americas which meant that indigenous peoples who were pressed into labour there typically died in short order. It was soon discovered that West Africans were unusually resilient in the face of malaria—more resilient than Europeans. They appeared to possess something which, thanks to Covid, we have become accustomed to call 'immunological dark matter'. [5] Now, obviously, malaria didn't *cause* slavery, but it was a powerful secondary shaping force on its history.

It seems safe to predict that Covid-19 will have a much smaller impact on world civilisation than the Columbian exchange! I begin with it nevertheless because it draws attention to something that is too often overlooked in accounts of the pandemic. Pandemics have far-reaching and often hidden cultural effects.

When Covid-19 first appeared in the West, it was unlike any disease we had known. It was both medical and 'more than medical'. The effects of the virus seemed to have more in common with a natural disaster or a terrorist outrage. For the first time since World War Two, mass graves were dug in Western Europe. Hospitals were overwhelmed by the sheer number of very sick patients who presented in the Spring of 2020. The scenes relayed on our television screens were on a par with what might be expected following a sarin gas attack. Family and other loved ones were required to put on personal protective equipment that resembled astronauts' space suits to be with the sick as they died. One wondered if the dying person even recognised their friends and relatives in this gear. Lockdown, likewise, amplified the sense of a health emergency and at the same time invited comparison with a different kind of historical catastrophe. In Europe, the only comparable precursor events in recent memory were the curfews imposed on civilian populations in wartime. The first lockdown was likened to the Siege of Sarajevo (1992–1995). In fact, the population of Sarajevo was studied intensively during the first lockdown to see if it held up more effectively than other communities under these unusual circumstances [6].

At the same time, the pandemic also created a sense of community, including a sense of online community. The fact that initially everyone was deemed to be at risk produced social solidarity of a kind not seen for many years. Governments nationalised large swathes of the economy, albeit temporarily. And because Covid-19 turned out to be a seasonal virus, infection rates fell sharply in the Summer of 2020. It seemed we were winning the war against Covid-19. Perhaps the enemy could be triumphed over and forgotten?

Scientists have not been slow to point out that the ecological release caused by Covid-19 has been beneficial to many other forms of life. In a powerful reminder that human activity is often a destructive influence on the environment, climate and ecology of the earth, the distinguished television naturalist David Attenborough made a programme for Apple TV documenting the benefits of this year of lockdowns to whales, cheetahs, monkeys, birds and deer, among many other species in various parts of the globe [7]. Many others have recorded improvements in the quality of air and water in some of the poorest and most polluted parts of our planet [8].

One way to think about events such as these is to observe that all living things require niches. A niche is an advantageous situation in any ecosystem. The evolutionary biologist Richard Lewontin introduced a scientific version of the concept in 1983 when he observed that 'organisms do not adapt to their environments; they construct them out of the bits and pieces of the external world' [9]. Humans, obviously, are the most successful architects of niches our planet has ever seen. We have reshaped the world in accordance with our own needs and wishes, snatching victories over nature in the form of industry, agriculture, science, to name but three. But as the Princeton bio-anthropologist Agustín Fuentes points out, nature always fights back. Our use of fossil fuels accelerates global warming, driving populations out of previously inhabitable areas. Intensive agriculture exhausts the soil. Our niches become less hospitable. And when this happens, it affords opportunities to other living organisms to create evolutionary niches more suited to their requirements. Covid is one such organism.

According to Fuentes, Covid was made possible by "the contemporary human practice of extracting wild animals from forested and other nonurban landscapes and bringing them into dense urban markets for processing and sale. It is this context of capture, confinement, crowding, butchering and ultimately consumption by humans that presents the opportunity for an exceptional virus with a few particular mutations to find a new host. These are rare events, and it is even rarer for them to result in the perfect storm of something like SARS- CoV-2 and the subsequent COVID-19" [10].

It is impossible to imagine Covid spreading at the rate it did without the assistance of our hyperconnected world. We live more densely than at any time in our history. Material inequalities, though on the decline globally, are still acute at national level in most countries in the rich world and, as the situations in the United States of America and Brazil especially show, political division has been a key driver in the spread of the virus.

Part II

One year into the pandemic, a peculiar situation obtains. Several effective vaccines have been trialled, licensed and employed. Real improvements have been made in the treatment of the virus. Dexamethasone, to take but one example, has been shown to reduce mortality in hospitalised patients [11]. Remdesivir has proved effective among adults who were hospitalized with Covid-19 and had evidence of lower respiratory tract infection [12]. And, *pace* Donald Trump, hydroxychloroquine turned out to be useless. We now have more precise knowledge of variables such as the infection fatality rate (the proportion of infected people who will die of the virus), the case fatality rate (the proportion of *symptomatic* infected people who die of it), and the reproduction number. According to Nicholas Christakis, it's a little under twice as infectious as most flus but only a sixth as infectious as measles [13]. It is the second-most deadly virus to have emerged in the last hundred years, killing 2.9 million people worldwide in the space of just a year. This puts it ahead of AIDS. Most importantly of all, we know much more about who is most at risk from the virus and why.

But the status of patients in this research is somewhat ambiguous. Are they to be helpers in the fight against the virus or can they become coproducers? A great deal is at stake in this question.

The fact that Covid was a mass phenomenon deemed to affect everybody or almost everybody meant that Covid was studied differently from other diseases. At King's College London, the Zoe Covid Symptom Study was set up under the direction of Professor Tim Spector [14]. Everyone over 18 in the UK was invited to participate in the study. After completing a short questionnaire, over four million recruits were asked to report their physical state each day over a mobile phone app, C-19, whether they were symptomatic or not. The Covid Symptom Study had the notable effect of recording the loss of smell and taste as a symptom before the World Health Organisation recognised it. There were many similar studies carried out in other countries.

Patient voices have been incorporated into the design and development of many large-scale studies of Covid on a scale seldom seen in other conditions. In the late Spring of 2020, patient voices even appeared to overtake those of biomedicine when Elisa Perego, an Italian researcher, coined the term 'Long Covid'. It has been called 'first illness to be made through patients finding one another on Twitter and other social media' [15].

There is still no generally agreed definition of Long Covid. The UK National Institute of Health and Care Excellence, the UK government body charged with improving outcomes for people using the NHS and other public health and social care services, sees 'post-Covid 19' syndrome occurring when 'signs and symptoms that develop during or after an infection consistent with Covid-19, continue for more than 12 weeks and are not explained by an alternative diagnosis' [16]. Other authorities use a 4-week cut-off. Many sufferers of Long Covid report symptoms lasting more than a year. The range of symptoms is also subject to wide variation. Chronic fatigue, 'brain fog', headaches, prolonged respiratory difficulties and chest pains have all been recorded as part of the post-viral syndrome associated with Covid-19. It is currently diagnosed using the most common symptoms as index symptoms but increasingly patients seem to be reporting *clusters* of symptoms, suggesting Long Covid may in fact not be one condition but several. It often has a remitting-relapsing pattern of presentation. One puzzle is that people who seem most at risk of Long Covid are demographically distinct from those who are at greatest risk of death in the acute phase of the disease. In acute Covid-19, the old are more at risk than the young, men more than women, and people of colour more than Caucasians. Cases of Long Covid are concentrated in 'early to middle' middle age groups, women seem more susceptible than men, and black and ethnic minority populations do not yet seem to be at higher risk [17].

Part III

It forms no part of my case to understate the achievements of biomedical research or its future utility. Biomedicine has already proved its worth many times over in the fight against Covid-19 and it will surely make further important discoveries in the understanding of Long Covid. But there is a danger that too exclusive a focus on known or conjectured risk factors, mechanisms, and pathways will deprive people whose experiences do not map so readily onto the findings of laboratory science of their proper share of epistemic authority [18]. If that seems far-fetched, think of what happens to people with long term glandular fever, or myoencephalitis (ME), or chronic fatigue syndrome (CFS). They are regularly thought to be malingering or to suffer from an undiagnosed psychiatric disorder. At this moment, in Western Europe at least, the demographic profile of the typical Long Covid sufferer overlaps considerably with patients with these other conditions: more female, younger, and, assuming whiteness is a proxy for privilege, whiter—people who are apt to be written off as 'the worried well'. As Callard and Perego have observed,

> Discrimination–including racism, sexism, ableism–helps explain why patients from marginalized/minoritized communities, many of whom are central to making Long Covid, have been denied platforms, and sometimes have decided not to place themselves in the spotlight to discuss a disease that compounds discrimination [15].

In one of the founding texts of the narrative medicine movement, *The Discourse of Medicine: Dialectics of Medical Interviews* (1985), the Harvard social psychologist Elliot G. Mishler argued that medical consultations were fundamentally sites of struggle between what he called the voice of biomedicine and the voice of the lifeworld [19] Mishler transcribed and analysed a number of medical interviews and found that the categories of biomedicine are naturalised in them while the categories coming from the patient's own life—his or her given world—are treated as specious. The physician gets to name the world of relevant facts and deviations from these are marked out as such through the physician's 'selective inattention.' The voice of biomedicine seeks to minimise the voice of the lifeworld as if it were completely free of context. Patients who refuse to play this language game may then be stigmatised. Mishler was articulating a point of view that a number of Harvard anthropologists were to build on in the early 1990s. The best theoretical statement I know of the 'Harvard position' is to be found in Byron Good's book *Medicine, Experience and Rationality* (1993). Good observed that in Western culture, disease is seen as 'paradigmatically biological' [20]. He went on: 'It takes a strong act of consciousness to denaturalize disease and contemplate it as a cultural domain.' Illness meanings and experience tend to be marginalised, as if people experienced their conditions exclusively through the lens of biomedicine, with little or no reference to culture. The social, psychological and behavioural dimensions of illness were assumed to be secondary, following the disease like the bow wave of a ship. Deep down, Western patients are assumed to 'know' that only somatic referents count. Any other responses they might bring to the illness experience belong to the realm of private metaphysical fantasy.

For at least three reasons, this approach is profoundly misleading. First, as Good points out, patients and their significant others relate to their conditions in ways that may appear to have nothing to do with the diagnosis. These 'subaltern' ways of experiencing illness often have a profound impact on how the onset and course of the disease is lived both by the patient and those he comes into contact with. The kinds of interviews Good and his Harvard colleagues Leon Eisenberg, Arthur Kleinman and Mary-Jo Delvecchio Good carried out with patients and carers in the

1990s and early 2000s used narrative theory to highlight the importance of these 'subaltern' ways of relating to the condition [21]. As Kleinman put it in his classic book, *The Illness Narratives* (1988), 'the meanings of chronic illness are created by the sick person and his or her circle to make over a wild, disordered *natural* occurrence into a more or less domesticated, mythologized, ritually controlled, therefore *cultural* experience' [22]. Second, Good notes, as Marshall Sahlins had before him, that imagining that people experience their illness exclusively or even mainly through the lens of biomedical belief entails seeing culture in terms of adaptation, and gives analytic primacy to 'the rational, autonomous care-seeker' [20, 23]. It is an economic conception of the individual who supposedly pursues maximum utility value (in this instance health) at any price. And yet study after study of how people actually behave in relation to health suggests otherwise [24]. Lastly, we now know, as Good - writing in 1994 - could not, that ordinary life adversity affects health in far-reaching ways. Ongoing research into the social determinants of health and in epigenetics suggest that psychological experience can actually change a person's biology. The idea that the psychological and social always play second fiddle to the biological does not frame epigenetic research on the long-term consequences of childhood trauma. Space does not permit me to describe the most important findings from this field, but it is clear that health research in the broadest sense has moved far beyond somatic referents [25].

As the Harvard group evolved, they moved away from the distinction between the natural and the cultural and between disease and illness and started to see biology as cultural 'through and through' [26]. *Medicine, Experience and Rationality* was the first major work in which this new understanding was operationalised. Henceforth, they would present medicine as—in Good's words—'one of the symbolic forms through which reality is formulated and organized in a distinctive manner'. Much of *Medicine, Experience and Rationality* is devoted to an analysis of how in the West biomedicine displaces every other object of interest from view in case of illness. Medical students are taught to construct reality by formulating their patients as medical projects. The only pertinent story to be told about illness is the one told by biomedicine.

What bound the Harvard group together was a shared commitment to methodological neutrality with respect to why patients seek care at all. In their view, the supremely relevant fact is that medical consultations 'create' their objects, whether these take place in the tent of an itinerant healer in rural Azerbaijan or in one of the teaching hospitals linked to Harvard University. This does not mean they have no foundations in biology. Rather, it means that in both cases, the objects created are social or cultural as much as biological. They are, in Bruno Latour's terms, 'hybrid entities' that biomedicine 'black-boxes' as if they were primarily biological, and social only in the secondary, derivative sense that they arise in particular contexts. (Blackboxing' is Latour's own neologism. It refers to 'the way scientific and technical work is made invisible by its own success. When a machine runs efficiently, when a matter of fact is settled, one need focus only on its inputs and outputs and not on its internal complexity. Thus, paradoxically, the more science and technology succeed, the more opaque and obscure they become') [27]. But biomedicine is first and foremost an instrument of seeing, and a very powerful one at that. It enables us to see, understand, or reveal things *differently*. At the same time, it throws the life-world under eclipse. In particular, it pays scant attention to things that cannot be captured histologically (through tissue analysis). Biomedicine hypostasizes particular biological conjunctions. It is bad at capturing the developmental and the dynamic. The consequence is a set of overvalued and somewhat static concepts that the neuroscientist-philosopher Iain McGilchrist describes as disembodied, abstracted, and isolated from context in which 'whatever depends on the implicit or can't be brought into focus and fixed, ceases to exist.' The meanings that are so crucial to the lived experience of illness thus fall away [28].

What has any of this got to do with Long Covid? Quite a lot! At the moment—I am writing in the Spring of 2021– the Covid-19 pandemic including Long Covid is being narrated almost exclusively in terms of the behaviour of the virus and especially its capacity for mutation. The only discipline to approach its relevance in public discourse is economics. (Two Harvard economists, David Cutler and Larry Summers have called it 'the $16tn virus'. They describe it as 'the greatest threat to prosperity and well-being the US has encountered since the Great Depression' [29]). In order to secure legitimacy for the concept of Long Covid, activists have to justify it in narrow biomedical terms. When the lifeworld appears at all, it tends to do so in ways that are consonant with a narrow biomedical point of view. We are told, for instance, very plausibly, that healthcare professionals are more likely to incur higher viral loads because of repeated exposures. The idea that Long Covid might also have far-reaching experiential precursors in childhood and in group life is being lost. What may also be lost is the possibility of revisioning illness itself and with it the sick person. I think it would be entirely in keeping with the Harvard approach to consider the patient presenting with Long Covid in Latourean terms as a node in a potentially infinite actor network that can, and indeed must, be studied in complex, cross-disciplinary ways. As Gregory Bateson pointed out more than half a century ago, the individual person is just one of the ways of perceiving differences in nature [30].

So how can we use the legacy of Covid-19 to inaugurate a new relationship with illness? There is no easy answer to this question. The power of illness to cause dread in a secular age that valorises youth and health is formidable and stymies constructive actions of all kinds. But here, by way of conclusion, are three modest proposals with that aim in view.

Take a Wide View of Any Aspect of the Formation of Long Covid

The actions biomedical theory is interested in take the form of linkages between biochemical objects. When psychosocial processes are left out of account, they disappear from view and often fail to achieve even secondary relevance. The remedy for this is to make as much room as possible for the voice of the lifeworld, as Mishler and Good advocated. Pay attention to the context. Covid has been very unusual amongst physical diseases in the sheer number of patient accounts that have been recognised to date. Studies like Callard's and Perego's represent an important beginning from which more complex analyses can be developed [15]. In this next stage, researchers would seek to take account of a much broader range of variables and processes than we are accustomed to do. As Rose and his colleagues have commented, they would recognise the distinct neurological, ecological and social pathways and mechanisms that shape human susceptibility [31].

Embrace Heterogeneity

It is already clear that Long Covid in particular takes many forms. Professor Nishi Chaturvedi of University College London has suggested that the syndrome is characterised by 'at least 50 symptoms'. It would be a pity if this list was whittled down prematurely because our models can only explain some of them. And we should study the symptoms in a variety of ways. Some patients with Long Covid have reported that their symptoms remitted following vaccination. That is cause for celebration. A wide variety of interventions should be studied, and we should widen our sense of what constitutes an intervention. In most illnesses like Long Covid, patients have to cope with a great deal of uncertainty. They don't really know what situation they are dealing with or how others will respond to their plight. Yet we all have to deal with uncertainty some of the time. We try to make sense of the situation we are in through trial and error. We should focus on the totality of the resources people draw upon from their lifeworlds to make the illness bearable.

Facilitate Cross-Cultural Comparisons, Giving Full Attention to the Experience of the Developing World

As I write this chapter, India is experiencing its 'second wave' of Covid-19. In New Delhi, hospitals are at breaking point, with patients sharing beds while others are turned away. One in three tests are positive. Given that in the West, Long Covid seems to be concentrated in 'early to middle' middle age groups, women seem more susceptible than men, and black and ethnic minority populations do not yet seem to be at higher risk of developing it, it will be interesting to see if Long Covid takes different forms there and afflicts a different demographic.

Conclusion

This chapter has explored the cultural evolution of Covid in the West. It began by highlighting the fact that major biological events always have far-reaching cultural consequences before homing in on the peculiarity central to this pandemic, namely, the part ordinary sufferers have played in understanding it. I suggested that Long Covid represents a triumph of patient involvement. It is an innovation that could potentially alter the way illness is seen in the rich world by highlighting the role of the life-world in sustaining it. To see how this might be done, I rehearsed some of the key arguments put forward by members of the Harvard medical anthropological school in the 1990s and 2000s. I concluded by making some brief proposals to keep research on Long Covid open.

Practice Time

1. In your opinion, why did Covid-19 produce so much social solidarity in the early stages of the pandemic? What happened to it as the pandemic unfolded?
2. What thoughts do you have about why Long Covid seems to affect a different demographic from Covid-19?
3. Could Long Covid change the social status of myoencephalitis (ME), or chronic fatigue syndrome (CFS)?
4. Who is best placed to help patients bring the voice of the lifeworld into the experience of Long Covid? Clinicians? Which ones? Family and friends?
5. Will Covid be in any paradigmatic in the public understanding of (a) disease and (b) public health?

References

1. Crosby AW. The Columbian exchange: biological and cultural consequences of 1492, 30th anniversary edition. Praeger; 2003.
2. Mann C. 1493: uncovering the new world Columbus created. New York, NY: Alfred A. Knopf; 2011.
3. Nunn N, Qian N. The Columbian exchange: a history of disease, food, and ideas. J Econ Perspect. 2010;24(2):163–88.
4. Crosby AW. Canto classics: ecological imperialism: the biological expansion of Europe, 900–1900. 2nd ed. Cambridge, UK: Cambridge University Press; 2015.
5. Friston K, Pillay D, Costello A. 'Dark matter', second waves and epidemiological modelling. BMJ Global Health [Internet]. 2020;5 https://doi.org/10.1136/bmjgh-2020-003978.
6. Reliving Sarajevo's siege under lockdown [Internet]. Iwpr.net. [cited 2021 Jun 5]. Available from: https://iwpr.net/global-voices/reliving-sarajevos-siege-under-lockdown.
7. Apple TV+ debuts "the year earth changed" to herald earth day 2021 [Internet]. Apple. com. [cited 2021 Jun 5]. Available from: https://www.apple.com/uk/newsroom/2021/03/apple-tv-plus-debuts-the-year-earth-changed-to-herald-earth-day-2021/.
8. Rodó X, San-José A, Kirchgatter K, López L. Changing climate and the COVID-19 pandemic: more than just heads or tails. Nat Med. 2021;576–9.
9. Lewontin R. Gene, organism and environment. In: Bendall DS, editor. Evolution from molecules to men. Cambridge: Cambridge University Press; 1986.
10. Fuentes A. A (Bio)anthropological view of the COVID-19 era midstream: beyond the infection. Anthropol Now. 2020;12(1):12–24.
11. RECOVERY Collaborative Group, Horby P, Lim WS, Emberson JR, Mafham M, Bell JL, et al. Dexamethasone in hospitalized patients with Covid-19. N Engl J Med. 2021;384(8):693–704.
12. Beigel JH, Tomashek KM, Dodd LE, Mehta AK, Zingman BS, Kalil AC, et al. Remdesivir for the treatment of Covid-19 - final report. N Engl J Med. 2020;383(19):1813–26.

13. Christakis NA. Apollo's arrow: the profound and enduring impact of Coronavirus on the way we live. London: Little, Brown Spark; 2020.
14. https://covid.joinzoe.com.
15. Callard F, Perego E. How and why patients made Long Covid. Soc Sci Med. 2021;268:113426.
16. https://www.nice.org.uk/guidance/ng188.
17. Sudre CH, Murray B, Varsavsky T, Graham MS, Penfold RS, Bowyer RC, et al. Attributes and predictors of long COVID. Nat Med. 2021;27(4):626–31.
18. Fricker M. Epistemic injustice: power and the ethics of knowing. Oxford: Oxford University Press; 2009.
19. Mishler EG. The discourse of medicine: dialectics of medical interviews. Westport, CT: Praeger; 1985.
20. Good BJ. Lewis Henry Morgan lectures: medicine, rationality and experience: an anthropological perspective. Cambridge: Cambridge University Press; 1993.
21. Good M-JD, Brodwin PE, Good JB, Kleinman A. Pain as human experience: an anthropological perspective. Berkeley, CA: University of California Press; 1992.
22. Kleinman A. The illness narratives: suffering, healing, and the human condition. London: Basic Books; 2020.
23. Sahlins M. Culture and practical reason. Chicago, IL: University of Chicago Press; 1978.
24. Scott S. Delay in seeking help. In: Llewellyn CD, Ayers S, McManus C, Newman S, Petrie KJ, Revenson TA, et al., editors. Cambridge handbook of psychology, health and medicine. 3rd ed. Cambridge: Cambridge University Press; 2019.
25. Bolton D, Gillett G. The biopsychosocial model of health and disease: new philosophical and scientific developments. Cham: Springer; 2019.
26. Kleinman A. The soul of care: the moral education of a husband and a doctor. London: Viking; 2019.
27. Latour B. Pandora's hope: essays on the reality of science studies. London: Harvard University Press; 1999.
28. McGilchrist I. The master and his emissary: the divided brain and the making of the western world. New Haven, CT: Yale University Press; 2019.
29. Cutler DM, Summers LH. The COVID-19 pandemic and the $16 trillion virus. JAMA. 2020;324(15):1495 6.
30. Bateson G. Steps to an ecology of mind: collected essays in anthropology, psychiatry, evolution and epistemology. Chicago, IL: University of Chicago Press; 2000.
31. Rose N, Birk R, Manning N. Towards neuroecosociality: mental health in adversity. Theory Cult Soc. 2021; https://doi.org/10.1177/0263276420981614.

Looking Behind the Veil: Why Narrative Medicine Matters in Times of Uncertainties

Susana Magalhães

Looking Behind the Veil

"In their emphasis on ways of looking at morality, Santayana and Temkin provide us with two conceptions which merge in medicine and literature and which ground their natural affinity for each other. For both must start by seeing life bare, without averting their gaze. Yet, neither can rest in mere looking. To be authentic they must look feelingly—with compassion. Medicine without compassion is mere technicism—curing without healing: literature without feeling is mere reporting—experience without meaning. (...) It is easy to forget that the physician's major diagnostic tool, despite the burgeoning of electrical, chemical and radiological techniques, remains the clinical history. That history is nothing more than a story, a complex, personal variegated story with multiple interlacing themes, in the life of a human being. Each story is unique. And as Peguy says, "When a man dies, he dies not just of the disease he has but of his whole life" [1].

Covid-19 has been building an uncanny reality around us. The invisible threat has been unveiling the fragility that medicine, science and western societies, hand in hand, have been managing to cover, to mask, to transfigure. Aging, illness, errors and mistakes and the *final flaw* –death—have been steadily translated as persistent youth, expected life-long healthy bodies, error-free actions and *death-less* lives. The pandemic has been reminding us that not only are we still vulnerable, but also that it is precisely this vulnerability that makes us human. Human enough to be caregivers, to risk our health to take care of those who fall sick and not to let panic rule our lives (this is a hard mission in a globalized world, where information flows every minute, drawing false images together with real ones, delivering fake news side by side with true stories).

S. Magalhães (✉)
Unit for Responsible Conduct in Research, Institute for Research and Innovation in Health, University of Porto, Porto, Portugal

M. G. Marini, J. McFarland (eds.), *Health Humanities for Quality of Care in Times of COVID -19*, New Paradigms in Healthcare, https://doi.org/10.1007/978-3-030-93359-3_3

The main individual ideal to be pursued in our days—freedom—is now presented with the limitations it has always had but few people were and still are willing to accept: the limits imposed by the common good and by the Other. The duty to care and community-wide quarantines are ethical and political issues that the whole world has been dealing with. Faced with the urgency of ethical deliberation and decision making, we resort to bioethical experts, philosophers, artists, in search of other news rather than the ones spread uninterruptedly by the mass media. This search for other narrative sources reminds us of the role of Bioethics at the beginning of its own story, when it was shaped by Van Potter as the bridge to the future: "Mankind is urgently in need of a new wisdom that will provide knowledge of how to use knowledge for man's survival and for improvement in the quality of life" [2]. Bioethics is therefore defined by Potter as the new wisdom rooted in biological knowledge and human values, that aims at providing an ethical frame to science and technology. The meta-knowledge envisaged by Potter provides alternative ways to applying scientific and technological advances, considering not only the here and now, but also the generations to be born. As applied ethics, Bioethics has a public dimension that has a strong impact on public opinion, providing the citizens with information and formation about the ethical issues underlying different areas of human action. The wide range of resources available in bioethics education provides different tools that can be explored according to the topic under discussion, the subjects participating in the ethical deliberation procedure and the kind of approach used to analyse the ethical issues. Using narratives as a resource for ethical deliberation means that this approach must be anthropological, bio-cultural, a revision of ethics as a dialogical search and not as a source of ready-made answers or recipes:

> Today logocentrical habits prevail, together with a philosophy of education focused on conveying information and training experts, overestimating the pragmatic and the scientific, the efficient and the instrumental, and disregarding the key role played by imagination and memory, which were the basis of the free, critical and creative thought of Greek civilization ([3], our translation).

In the current pandemic situation, not only are we faced with the unpredictable and the uncontrollable, but we are also required to plan ahead, to learn with the events, identifying the resources that have been used, those that were not available and need to be provided/discovered and above all we need to revisit terms and concepts in the context of healthcare. Perceiving health as relational means to set up connections among all stakeholders and recognize the impact of their interactions on each one of them. Re-thinking the meaning of care as healing integrates personhood into patienthood and broadens time into the past before the pandemic, the comprised present and the possibilities inscribed in the future. Such dialogical revision of terms and concepts underlines the central role of language not only as a linguistic tool but also as a conceptual one that shapes how we behave and act. Assuming that we actually live in subjective worlds that are simultaneously objective to each of us and that multiplicity is in fact the very essence of human experience, then we can grasp the important role of narratives as a source of phenomenological truth, i.e. as the place where the objective truth of the external world and the subjective truth of the individual meet:

> Metaphor is living not only to the extent that it vivifies a constituted language. Metaphor is living by virtue of the fact that it introduces the spark of imagination into a "thinking more" at the conceptual level. The struggle to "think more", guided by the "vivifying principle", is the "soul" of interpretation. [4, p. 303]

During this pandemic, the metaphor of war has enacted a survival mood, making us perceive the virus as the enemy and the government measures as exceptional ones justified by this condition.But this particular metaphor has also widened up its domain, including the Other as the enemy and health professionals as soldiers at the frontline, ready to kill, and the hospital as the battleground. Being aware of the central role of language as shaping not only the way we perceive the world but also the way we act, reminds us that actually *the ends of medicine derive from the goals of human life, and these are primarily social and political.* The dizzying advances in technology and the profound socio-political changes in a world caught up in globalization and *multiculturalism* stress the Socratic need to know yourself, i.e. know your limits. The more scientific and technological progress advance, the more powerful and the more vulnerable do we become. In fact, vulnerability as potential wound, as probability of getting harmed, is strictly related both to solicitude and power, since being solicitous as well as being powerful places the individual on the fringe of vulnerability. This pandemic has revealed that we are all vulnerable, but still there are those who are more vulnerable than others due to previous vulnerabilities, thus imposing the ethical duty to protect them, setting up a link between solicitude and vulnerability. Such an awareness requires actively listening to patients, health professionals, caregivers and integrating more than facts and news into healthcare. Instead of the war metaphor, there has emerged the need to resort to the mirror metaphor, so that we can see ourselves in the eyes of the Other. Together with change of metaphor, we have been experiencing what John Keats defined as the *negative capability*, i.e. being able to live with uncertainties, mysteries, the unknown,[1] within a web of connections that usually escape logical thinking: "In addition to discerning what sort of disease or problem a person may have, GPs also need to know something of the sort of person a disease, concern, or a symptom may have. Narrative sense of these developing relationships is fundamental to the care and clinical management of individuals over time" [5].

Ethical issues are always embedded in time, space, language, enacted by human beings, voiced by different characters, therefore they are intrinsically narrative. Narrative medicine provides tools and concepts to address the narrativity of ethical issues within healthcare, while allowing patients, caregivers and health

---

[1] **On Negative Capability: Letter to George and Tom Keats,** Hampstead, Sunday, 22 December 1818: "at once it struck me, what quality went to form a Man of Achievement especially in Literature & which Shakespeare possessed so enormously—I mean *Negative Capability*, that is when man is capable of being in uncertainties, Mysteries, doubts, without any irritable reaching after fact & reason (…)."

professionals to see themselves in the mirror, recognizing their own vulnerabilities. It is from this recognition that the capacity to respond to the other's suffering can flourish, connecting biology and biography, past, present and future:

> With narrative competence, physicians can reach and join their patients in illness, recognize their own personal journeys through medicine, acknowledge kinship with and duties toward other health care professionals, and inaugurate consequential discourse with the public about health care. By bridging the divides that separate physicians from patients, themselves, colleagues, and society, narrative medicine offers fresh opportunities for respectful, empathic, and nourishing medical care [6].

Narrative Medicine and Bioethics

According to Rita Charon, Narrative Medicine is "practised with the narrative competency to recognize, interpret, and be moved to action by the predicament of others" [7]. Brian Hurwitz considers that "Narrative Medicine is a practice and an intellectual stance which enables physicians to look beyond the biological mechanisms at the centre of conventional approaches to medical practice, towards domains of thought and ways of telling that focus on language and representation, on the emotions and relationships which illuminate health care practice" [8]. Narrative Medicine main focus is not to collect more data about the patient, but to set up relationships within the healthcare environment that allow for the integration of personal data in patients care, as well as in team work and in individual practice. It is not only about gathering more data; it is about questioning the nature of knowledge and how knowledge gains legitimacy. So symptoms should be situated within the stories patients weave to make sense of their lives. The perception of clinical knowledge as co-constructed has led to new platforms of patient-doctor communication based on sharing the clinical record and co-building it. Instead of a one size fits all approach, narrative categories like voice, character, time, space and plot remind us that evidence should also be practice and individual-based. Only then can care delivery, design and management be actually improved. All stakeholders are involved in this narrative approach—patients, relatives/caregivers and health team. It presents the same kind of challenge that an evidence-based approach does: knowing how and when to stop.

As Maria Giulia Marini reminds us "symbiosis between Evidence Based Medicine and Narrative Medicine should be the best achievement, a cohousing made by sharing technologies, arts and competences" [9]. Moreover, Marini points out that Narrative Medicine is democratic, being able to connect patients and healthcare providers. The reality of a disease is always an object of interpretation, whether we are talking about the objective interpretation of a scientist, an evidence-based physician or the subjective interpretation of the person who suffers and tries to communicate suffering to another. It can also be the subjective interpretation of the person who cares (physician, nurse, caregiver) or the subjective interpretation of society based upon shared pre-judices, cognitive biases, moral values, cultural beliefs. Evidence-based medicine, can hardly integrate the intersubjective dimension of therapeutic care, leaving both patients and health professionals deprived of their personal needs, because as Eric Cassel points out "each of us gets to our illness our own way, it becomes part of our story, and we individualize it by its place in the narrative of our lives. To know that illness we must know something of the person. To know the person, we must know something of the narrative" [10].

Patient- centred care—based on the principles of dignity, compassion and respect, with the aim to promote enabling, personalized and coordinated care—was designed as a response to the lack of the intersubjective dimension. However, in order to be fully implemented, it requires tools and reasoning skills that can actually provide thickness to what might just be abstract principles. Dignity, Autonomy and Vulnerability need to be filled with the depth of lived experience to be meaningful, and this can only be achieved if attitudes, behaviour, gestures and relationships are modelled by narrative skills. Such skills make a difference by fostering the capacity to integrate uncertainty and unpredictability into clinical reasoning. In fact, narrative requires openness to changes in the plot thread, curiosity, the ability to move backwards and forwards so that one can eventually give meaning to the whole story, and attention to the meaningful details. Within this narrative ecosystem,

a. Dignity, as Nordenfelt has argued in his study on Dignity and Older Europeans (Fifth Frame-work (Quality of Life) Programme) [11] has different dimensions, namely the dignity of merit, the dignity of moral or existential stature, the dignity of identity and the universal human dignity (Menschenwürde):

> 1) The Menschenwürde pertains to all human beings to the same extent and cannot be lost as long as the persons exist; 2) The dignity of merit depends on social rank and formal positions in life. There are many species of this kind of dignity, and it is very unevenly distributed among human beings. The dignity of merit exists in degrees, and it can come and go; 3) the dignity of moral stature is the result of the moral deeds of the subject; likewise, it can be reduced or lost through his or her immoral deeds. This kind of dignity is tied to the idea of a dignified character and of dignity as a virtue. The dignity of moral stature is a dignity of degree, and it is also unevenly distributed among humans; 4) The dignity of identity is tied to the integrity of the subject's body and mind, and in many instances, although not always, also dependent on the subject's self-image. This dignity can come and go as a result of the deeds of fellow human beings and also as a result of changes in the subject's body and mind [11].

b. Autonomy is perceived as a grading principle that can affect different capacities in one's life, mainly the capacity to make free and informed decisions; the capacity to narrate one's life in a coherent way; the capacity to perform one's daily routine activities; the capacity to execute what was decided before and maintain it throughout time; the capacity to access and control one's own private and public information:

> In recent years, several studies have advocated the need to expand the concept of patient autonomy beyond the capacity to deliberate and make decisions regarding a specific medical intervention or treatment (decision-making or decisional autonomy). Arguing along the same lines, this paper proposes a multidimensional concept of patient autonomy (decisional, executive, functional, informative, and narrative) and argues that determining the

specific aspect of autonomy affected is the first step toward protecting or promoting (and respecting) patient autonomy. These different manifestations of autonomy are not mutually dependent; there may be patients who have problems in one dimension, while at the same time being fully autonomous in others.(…) In this paper, I shall examine some of these interactions and show how they may lie at the heart of the problem of poor treatment adherence in many patients with chronic ailments (where adherence is defined as being the extent to which a patient's behaviour over time coincides with the recommendations made by and agreed with their health professional) [12].

Relational autonomy is a concept that has been issued by the need to find an adequate explanation of impairment of autonomy in contexts of oppressive socialization, together with feminist critiques of traditional notions of autonomy:

Crudely stated the charge is that the concept of autonomy is inherently masculinist, that it is inextricably bound up with masculine character ideals, with assumptions about selfhood and agency that are metaphysically, epistemologically, and ethically problematic from a feminist perspective, and with political traditions that historically have been hostile to women's interests and freedom. What lies at the heart of these charges is the conviction that the notion of individual autonomy is fundamentally individualistic and rationalistic [13].

Mackenzie and Stoljar believe that, although feminist critiques have pointed out serious problems with some historical and contemporary notions of autonomy, this concept should be repudiated altogether. In fact, these critical approaches have contributed to a reconceptualization of autonomy, and it is this refigured concept that is called relational autonomy. From a relational point of view, individuals are perceived as emotional, embodied, desiring, creative, and feeling, as well as rational, creatures" [13], and only by acknowledging these features, do we respect each individual's autonomy, the capacity to choose their course of action, as well as their own communities.

c. Vulnerability is framed in the plural as different layers [14] that can cover each other, increasing the complexity of one's health condition: cognitive, medical, juridic, deferential, infrastructural and allocational vulnerabilities [15] can actually superimpose on each other or they can exist independently. Considering these layers of vulnerability rather than a single label shifts the focus to what actually matters to each person and raises the likelihood that individualized care will be provided.

The narrative turn in healthcare is due to different factors: 1. The rise of chronic illnesses and their heterogeneous character has led to the loosening of the biomedical focus, providing space for lay narratives to flourish; 2. The high costs of high tech bio-medical care has shifted the emphasis from the hospital sector to primary care, where listening to patients within a holistic care approach provides an opportunity for patients narrative to be given attention; 3. A recent and more powerful democratic impulse has allowed lay people to insert themselves in healthcare as new carriers of credit; 4. The internet and alternative forms of healthcare provide lay people with an increasing range of information, thus reducing the role of medical authority as the source of all official knowledge about illness. The effects of an age-ing population and the predominance of chronic illness explain why management and care have superseded treatment and cure.

Therefore, narrative machinery is of utmost importance in healthcare for three reasons:

- Firstly, while biology is *still* here, we can say that we are complex, unique, sin-gular, non-replicable beings, so we can only be cared for if our physicians and caregivers remain persons able to see us in a singular way;
- Secondly, it is the narrative machinery that allows us to see what is essential for Patient A diagnosed with Parkinson, whose preferences, concerns and needs are different from patient B with the same diagnosis. These differences can be inter-preted and acted upon if health care professionals can distinguish between what Meza and Passerman [16] call disease-oriented evidence (DOES)—finding the right diagnosis and the clinically meaningful values, percentages and numbers, and patient-oriented evidence that matters (POEMS)—finding out what matters for this particular patient: is it working, being autonomous, taking care of grand-children, seeing, moving?
- Thirdly, narrative machinery is required if we want to save human and material resources.

The three reasons stated above underline the fact that care is a process and not a moment and it is about

a. What we do, how we do, what we say while we do
b. What we say
c. What we omit
d. What we distort/manipulate
e. How we attend to verbal and non-verbal communication signs
f. Integrating the question of HOW the *here and now* has been reached, together with the question of WHAT the diagnosis is.

    Acknowledging subjectivity means being able to face prejudice and how it may affect misdiagnosis and patient's compliance. Narrative skills –such as attention, interpretation, representation, and imagination lead to affiliation, thus precluding what Miranda Fricker [17] calls Epistemic Injustice, comprising two forms: testimonial injustice and hermeneutic injustice. The former takes place when the speaker's word is not given credibility due to prejudice, whereas the latter occurs when there is a gap in collective interpretative resources that pre-vents someone from making sense of their social experience. If physicians and other health professionals can actually shape what patients tell, by their posture, tone of voice, the words they say and those that remain unsaid or distorted, then communication strategies fall short of their target if they are just a list of what to say and how to say it. Although the focus has been on patient care, the role of narrative in medicine is perceived as much wider and deeper, reaching not only the quality of patient care but also the three Rs of good medical practice:

Relationship, Recognition of one's own and of others' Vulnerability, Reflection (or, we shall say, Responsibility, Respect, Reconciliation).

The relational approach of the so-called narrative medicine requires two main tasks from health professionals: relocating personhood within patienthood and professionalism by means of acknowledging the subjective elements at play at the clinical encounter; and, as already mentioned before, reconceptualizing terms and concepts as means to reconciliation with their own selves and with all those who take part in health care. From a narrative medicine approach, health is not only what a body has, but how a body relates to other bodies and to the environment, being thus perceived as a dynamic concept. Moreover, the body is both the lived body and the object body, which means that the body itself, which is the focus of interest for evidence-based medicine is always both subjective and objective in terms of its relation to the world and in terms of its place in the world: the body is not only in space but of space and the temporal structure of bodily experience always entangles past and present. Likewise, pain must be reconceptualized, bearing in mind that how we think about pain is ideally guided by those unspoken and unconscious assumptions, myths, and metaphors that shape our understanding of the individual sufferer's reality and experience of pain. "Commonly-utilized diagnostic tests cannot communicate the intensity, duration, quality, and personal dimensions of pain, as this is something that only the sufferer can appreciate" [18]. By emphasizing this first-person narrative, the entirety of the person in pain can be considered, rather than simply focusing on signs and symptoms.

Narrative Medicine is addressing these issues from a narrative point of view, which is aligned with the so-called Narrative Bioethics: "The only way of taking account of social phenomena is by means of communication. The person entangled in ethical dilemmas and the bioethicist who is theoretically interested in ethical problems do have to talk to each other. This is the first and most essential relation, which the two parties have to enter. We have to keep in our mind, however, that the project of a narrative bioethics is not in competition with traditional forms of ethical justification. The necessity of an alternative ethical concept results from the blind spots of an ethics based on principles alone" [19]. The ethical problems raised by the pandemic belong to different levels: (a) the micro level of decision making that affects one's own life; (b) the intermediate level regarding how health care professionals actually care for their patients and caregivers and how they relate to each other and to the system; (c) the macro level of governmental decisions that impose restrictions of movement and lifestyle changes on the citizens. Triage criteria, resources allocation inside and outside of hospitals, family dynamics—the ethical problems that have risen throughout this pandemic cannot be answered without considering the exceptional context, thus without resorting to images, metaphors and concepts that frame the decision making within a certain plot, with particular characters, ranging from victims to heroes. Behind the veil of an abstract plot, real persons have experienced and still experience pathological mournings, professional burnout, depression and other psychological disorders. These are the ones that remind us that there is no way out of the tragedy with decisions that actually save all the values at stake: autonomy, dignity, integrity, equity. There is one window that opens up to the future horizon: learning and doing differently, which requires listening to the narratives of all of us, paying special attention to those who have been particularly affected. Narrative Bioethics and Narrative Medicine provide us with this window and, like cautionary tales, they also make us entangle past, present and future into a meaningful narrative, where there is room for all those who had not been well cared for before the outbreak of the pandemic and have had their chronic diseases aggravated with the virus; all those who were being cared for before the pandemic and have been neglected since the virus started spreading inside their villages, cities, country; all those who lost the loved ones and have not mourned properly; all those who were pictured as heroes and were not cared for as human beings.

Pandemic Narratives and Art: Bridge to the Future

> By narrating a life of which I am not the author as to existence, I make myself its co-author as to its meaning (…) It is precisely because of the elusive character of real life that we need the help of fiction to organize life retrospectively, after the fact, prepared to take as provisional and open to revision any figure of emplotment borrowed from fiction or from history [20].

Art is also a pillar for our face to face with the community, the Greek polis, since "it can encourage the facility, the willingness and the ability to enter into the larger public debate in these cacophonous times, when collective silence does not serve. And, finally, it can also provide an "outlet for both emotional engagement and self-reflection in a culture that typically denies both, looks outward rather than inward, and too often ignores not only the personal costs but the personal rewards of medical work" [21]. The narratives of the pandemic, for example, are sources of information, upon which we can reflect (think with and not only think about) and learn with the pandemic. Art makes us see ourselves in the mirror, but always with an open window to an alternative world. Before the pandemic started, we were living in a palliative society, where pain and suffering were silenced. We were living a pain-free society, jumping from project to project. We forgot that pain purifies, it causes catharsis. Art was thus conceived as moral free, a decoration item, an anesthetized form of creativity. The virus has revealed to us the society we were living in, and Art can help us to find out not only who we were and are but also who we can be. Narratives of the pandemic reveal that we need to focus on good life and not only on survival and through Art we can reintroduce in our lives the pain, the suffering, the uncanny, the otherness we were repressing, silencing, erasing. Moreover, Art can make us experience and understand the meaning of mortality as the condition that frames the search for meaning as the first ethical duty of human beings. Interestingly, the forbidden hug of our pandemic time reminds us that this gesture—hugging—is the symbol of what technology has not invaded yet. As João Luís Barreto Guimarães says in one of his poems, *The Mechanics of a Hug*, "what you hold in a hug when you hug someone is not a body: it is time; but the strength of your hugs is weaker than that of time and it's you who have to let go, because time does not accept to be still for so long and requires you to release it so that it can return to movement" ([22], our translation). This pandemic makes us miss hugs and this missing reminds us of what actually makes us human: the capacity to think metaphorically, to see something beyond what that something is used for or beyond the attempt to close it in a strict definition. It is precisely the metaphorical imagination and creativity that provide us with new opportunities to apply the narrative medicine tools of close reading, reflective writing and parallel chart [23, 24]. Bearing in mind the need to listen to the person of the patient and the person of the professional in order to provide *Relationship Based Healthcare* and focusing on the ethical issues raised by the pandemic, we need to apply these tools in training, practice and research:

a. At Medical and other Health Sciences Schools, the curricula could integrate close reading and reflective practice since the beginning of the courses, with anatomy and semiology classes followed by reading and writing moments, inspired by books like *Adventures in Human Being*, by Gavin Francis as well as poems, short-stories, extracts from novels and real-life narratives;
b. Students could be required to observe (without intervening) emergency rooms, wards, ICUs, corridors of their hospitals;
c. Students and healthcare professionals could be invited to provide challenging ideas to change routines, procedures and logistics at hospitals and healthcare centres;
d. Healthcare professionals could be rewarded whenever they suggest new ideas to change ingrained habits that actually prevent good care;
e. Healthcare professionals could integrate reflective post its in their daily practice, not as reminders of *what to do or how to do*, but as reminders of issues related with the essential question of *who one is/who one wants to be:*

> The day I met error in my hospital;
> The day I realized that being a doctor (other health professional) can hurt;
> The patient who died to me;
> The day I had to provide care without empathy;
> The day I needed to be cared for;
> The day I could not take care of a patient;
> The day communication was broken (...)

These post-its could be discussed once a month in departmental meetings.

f. *Waiting Rooms* could be renamed as *Hospitality Rooms*, where patients and their relatives would experience waiting as productive and connective moments: notebooks (on paper or digital) could be available for writing/drawing about what they want to say to their doctor; literacy corners could be built around the room, so that technical terms could be translated into words that actually communicate what patients need to understand;

g. *Dead spaces* around the hospital could be renovated into living ones, enabling teenagers who are not bedridden to move around, listen to music, chat, do some physical exercise, *connect to life;*
h. Wards and ICU beds could have word clouds to identify the patients being taken care of, referring to *what matters to them* rather than *what their diagnosis and clinical record is.*
i. Architects could have open channels of communication with those using healthcare services so that the design of the buildings serves their purposes;
j. Terms and concepts could be reflected upon by health teams in order to define those that break communication, do harm to patients/relatives/health professionals, widen the gap between those being cared for and those taking care of;
k. Multidisciplinary healthcare teams would effectively coordinate their work, setting up a communications network that would not require the patient to answer the same questions over and over again or to cross the hospital back and forward to be examined by different professionals;
l. GPs could be integrated into the care of their patients when they are hospitalised, building coherent narratives that honour their stories.
m. *Digital Narrative Medicine Platforms* could strengthen the utility of technologies, bridging the gap between those afflicted by diseases and those who take care of them and bringing the focus to what really matters to each of them.

Considering that our identity is an inter-subjective process built throughout our life, and that narrative has a role in this process, it is possible to conceive of education as a hermeneutic task, linked with ethics since it is not possible to interpret without judging. This web, which connects ethics, hermeneutics, literary and real-life narratives and ethics, opens up new ways of seeing reality, of acting upon it and makes us suspicious of absolute certainties and closed narratives. According to Gracia [25] moral perfection can only be achieved through the ethical coherence between ideas and actions and in order to achieve that perfection this coherence must be deep and time-resistant. When this happens, our actions become virtuous habits that we enact effortlessly and even with some pleasure. Healthcare education and practice must therefore include tools and strategies that promote self-reflection, interpretation, insight, communication and narrative skills, if we want to make sure that this pandemic opens up a window to a better future, with time and space for *what we really care for.*

References

1. Pellegrino ED. To look feelingly: the affinities of medicine and literature. Lit Med. 1982;1(1):19–23.
2. Potter VR. Bioethics: bridge to the future. Englewood Cliffs: Prentice-Hall; 1971.
3. Clavel JM. Bioética y antropologia. Madrid: Publicaciones de la Universidad Pontificia Comillas; 2004.
4. Ricoeur P. The rule of metaphor: multi-disciplinary studies of the construction of meaning in language, trans. Robert Czerny with Kathleen McLaughlin and John Costello. Toronto and Buffalo: University of Toronto Press; 1977.

 5. Hurwitz B. Narrative and the practice of Medicine. Lancet. 2000;356(208):6–89.
 6. Charon R. Narrative medicine: a model for empathy, reflection, profession, and trust. JAMA. 2001;286(15):1897–902. https://doi.org/10.1001/jama.286.15.1897.
 7. Charon R. Narrative medicine: form, function, and ethics. Ann Intern Med. 2001;134:83–7. https://doi.org/10.7326/0003-4819-134-1-200101020-00024.
 8. Hurtwitz B. Narrative (in) medicine. In: Spinozzi P, Hurtwitz B, editors. Discourses and narrations in the biosciences. Göttingen: Vandenhoeck & Ruprecht Unipress; 2011. p. 73–8.
 9. Marini MG. Languages of care in narrative medicine: words, space and time in the healthcare ecosystem. Cham: Springer; 2019.
10. Cassel E. The Nature of suffering and the goals of medicine. Oxford: O.U.P; 1991.
11. Nordenfelt L. The varieties of dignity. Health Care Anal. 2004;12(2):69–81.
12. Arrieta Valero I. Autonomies in interaction: dimensions of patient autonomy and non-adherence to treatment. Front Psychol. 2019;14(10):1857. https://doi.org/10.3389/fpsyg.2019.01857.
13. Mackenzie C, Stoljar N, editors. Relational autonomy: feminist perspectives on autonomy, agency, and the social self. New York: Oxford University Press; 2009.
14. Luna F. Identifying and evaluating layers of vulnerability – a way forward. Dev World Bioeth. 2019;19(2):86–95.
15. Kipnis K. Vulnerability in research subjects: a bioethical taxonomy. in ethical and policy issues in research involving human research participants. Bethesda: National Bioethics Advisory Commission; 2001. p. 1–12.
16. Meza J, Passerman D. Integrating narrative medicine and evidence-based medicine: the everyday social practice of healing. London: Radcliffe Publishing; 2011.
17. Fricker M. Epistemic injustice: power and the ethics of knowing. Oxford University Press; 2007.
18. Carvalho AS, Martins Pereira S, Jácomo A, Magalhães S, Araújo J, Schatman ME, et al. Ethical decision making in pain management: a conceptual framework. J Pain Res. 2019;11:967–76. https://doi.org/10.2147/JPR.S162926.
19. Dubiel H. What is narrative bioethics. Front Integr Neurosci. 2011;5:10. https://doi.org/10.3389/fnint.2011.00010.
20. Ricouer P. transl. Kathleen Blamey. Oneself as Another. Chicago: Chicago University Press; 1992.
21. Huyler F. The woman in the mirror: humanities in medicine. Acad Med. 2013;88:918–20.
22. Barreto Guimarães JL. Nómada. Lisboa: Quetzal-Editores; 2018.
23. Charon R, Hermann N, Devlin MJ. Close reading and creative writing in clinical education: teaching attention, representation, and affiliation. Acad Med. 2016;91(3):345–50. https://doi.org/10.1097/ACM.0000000000000827.
24. Charon R. "Narrative reciprocity," narrative ethics: the role of stories in bioethics, special report. Hastings Cent Rep. 2014;44(1):S21–4. https://doi.org/10.1002/hast.264.
25. Gracia D. Fundamentos de Bioética. Coim.

The Consolation of the Written Word: Reading to Engage and Escape Our Pandemic Year

4

Carol-Ann Farkas

Human beings are very bad at predicting the future. When asked to envision what we think will make us happy and fulfilled, or what will crush us utterly, we consistently guess wrong. We're completely lacking in objective rationality when it comes to reconciling our (often fantastical) ideals about ourselves with our real behaviour. At the same time, we vastly underestimate our own resilience and adaptability [1].

So, as I write this in March 2021, a year after first being confronted with a deadly, global pandemic, it's hardly surprising to find that very little went as we expected it would. Maybe we imagined *then* that our pandemic year would be one of noble, ambitious, productivity…that perhaps never materialized. Maybe we believed *then* that we weren't sufficiently resourceful or resilient to cope…and *now* are quietly, pleasantly surprised that we coped after all.

The point is, that *then*, at the start of the pandemic, none of us had any idea of what we were doing.

That didn't stop us from making plans and predictions, though. One we all shared: *we were going to get so much done.* Back in March of 2020, I believed that I was going to use *all the extra time* freed up by the quarantine to do a *ton* of reading and writing. My friends—brilliant, creative, and often intimidatingly productive in their various *metiers*—all had similar visions for themselves.

The written word was going to be the lynchpin of our strategy for getting through it, even as we had no idea what *it* was.

This essay is… just that, an *essai*, an attempt, a meditation on the power of the written word to console and divert, during Our Year of Covid. I had intended this to be a very personal narrative… except, after a year of social distance,

C.-A. Farkas (✉)
BA Program in Health Humanities, School of Arts and Sciences,
MCPHS University, Boston, MA, USA
e-mail: carol-ann.farkas@mcphs.edu

© The Author(s), under exclusive license to Springer Nature
Switzerland AG 2022
M. G. Marini, J. McFarland (eds.), *Health Humanities for Quality of Care in Times of COVID -19*, New Paradigms in Healthcare,
https://doi.org/10.1007/978-3-030-93359-3_4

confinement, and an unlooked-for and unwelcome stasis, I'm as sick of the contents of my personal brain as I am bored to tears by the contents of my once-cherished, now-very-overrated personal space. What I need to do in this essay, at this point in the pandemic, is spare myself from myself, escape the confinement, and, figuratively if not yet literally, spend some time enjoying what we all miss right now more than anything—other people's company, other people's stories.

This project then, is an assemblage of experiences gathered from some of the most interesting and interested people I know: college professors, scientists, therapists, novelists, poets, athletes, contact tracers, architects, wood workers, sex workers, friends. To gather these responses, my methods were deliberately unmethodical, not so much research, as conversation, both spoken and epistolary. I invited my friends to talk or write, to tell me—and so, you: What was the pandemic like for them? In particular, what did they read, how, and how did it help—or in some cases, harm—as we all struggled to endure the past year?

Before I begin, it's important to recognize that the voices which I have gathered here can hardly be taken as representative of much, other than of a collective, undeniable privilege. Mostly white women, mostly "non-essential," able to work from home, the majority of us were protected not only from disease, but from the worst ravages of economic upheaval and eviction, and racial violence. Friends with small children at home, those who are non-white, those working in health care—all, for real, inexcusable reasons, were busy elsewhere. And there are voices that are present here only by their absence: those who have suffered with, and succumbed to Covid; those who have been stunned into silence by the exhaustion of caring for the sick, and of losing so many.

As I write this, we've all been marking the anniversary of the beginnings of the SARS-CoV-2 pandemic, more or less a year ago for those of us in North America. We all remember how *our* pandemic began, "the last good day" [2] amidst news reports and speculation and the initial plans which we took a little seriously but not terribly so because we figured—like the world in the summer of 1914—that this would all be over in just a few weeks. Now we look back at that time, aghast at our ignorance and denial.

You remember, don't you, how frightened we all were, and how disoriented? You remember how exhausting it was, not knowing if anything was safe, and so, having to assume that nothing was? Friends and strangers became equally dangerous, every object potentially, lethally, contaminated: park benches, groceries, our clothes, our hair. The familiar was made suddenly, incomprehensibly strange: hadn't I been chafing, throughout the winter, for a break from routine, for a chance to be excused from workouts and meetings, from subway cars crowded with hostile and indifferent strangers, from being beholden to other people's requirements of my time, my thoughts, my physical and mental energy? *Be careful what you wish for...* Now, all that time and peace and quiet should have been a gift, a comfort, and yet was also fraught, oppression and constraint and isolation masquerading as comfort and freedom (Fig. 4.1).

We were determined to make the best of it.

Fig. 4.1 Photo by
Carol-Ann Farkas

We posted about our intentions to "keep calm and carry on!"; we shared memes joking about introverts being the best prepared for quarantine life; we insisted that now was the time to take a break from the "attention economy" [3], that this enforced retreat from the world was not an ordeal, but an opportunity that we must welcome (gratitude journals, everyone!) and certainly not squander. So we set ourselves detailed schedules, designating blocks of time every day for mastering sourdough and meditation; committing to online classes in cooking, salsa dancing, and art history; finally writing that novel or memoir, knitting that sweater, getting back to singing lessons or oil painting or pointe shoes, or whatever aspects of our identities that had been pushed aside years ago in order to manage jobs and families but which we still believed were a part of us, things we were good at, disciplined at, liked. *We were going to get so much done.*

And above all, we were going to READ.

In building our expectations for a quarantine, we figured, initially, that it would just be a kind of glorified snow day—a bonus holiday, an excuse for idleness and having whisky-spiked cocoa in the middle of the afternoon. Like the snow day, this would be a temporary break, followed by some messiness as we all got back to the normal routine.

We were snowed in, all right, but not in the fun way. Aspiration and intentionality very quickly became buried beneath demands we weren't ready for. It wasn't easy to adapt to working, and working out, and socializing, and teaching our kids online—but we had to, so we did; the real struggle was the awareness that something was happening at the level of feeling, of mind—something big, and yet elusive to conscious perception and analysis. Why would it be otherwise? This was, as we all tired of saying very early on, *unprecedented*. But whatever it was, even if we were as safely removed and privileged as we could be, we felt the weight of it: the horrifying illness and death, the uncertainty, the incompetence and grotesque indifference of leadership, the isolation, conflicts with neighbours, friends, and family over wildly differing tolerance for risk, more uncertainty, isolation, a lot of anger, and not a little boredom.

Within a couple of weeks of the initial lockdown, the memes about sacralising this moment, dedicating it to compassionate mindfulness and creative growth had given way to memes reassuring one another that shattered concentration and exhaustion were (are) typical responses to trauma [4, 5]. For many, many reasons, including some we can't even fully understand because we're still in it, this past year has been *hard*.

The pandemic brought with it its own unlooked-for injuries, of course; it also made things that would have been challenging under "normal" circumstances, that much harder. For so many, the pandemic complicated life transitions and taxed emotional and economic reserves. Frank had just taken early retirement from a successful, but stressful, career as an academic researcher. He was looking forward to 2020 as the start of a new phase of his life, traveling, pursuing long-neglected interests. The pandemic had other ideas. But at least Frank was financially prepared to be stuck at home. Tom, in biotech, and Sarah[1], in fund-raising, had both been laid off at the end of 2019, and it was months before they could find work again as industries everywhere cut back on operations and froze hiring. Sunny, a personal trainer, and Kaycie, a sex worker, both relied, if for different reasons, on very hands-on relationships with clients paying by the hour; being cut off from their respective revenue streams has been a financial blow to both. Sarah and Anne, had been forced to relocate for work and had to struggle through this past year far from family, friends, or even a friendly culture—Sarah, a self-described urbanite, was stuck in a rural part of our state; Anne, very outspoken as a feminist, socialist, and anti-racist poet, felt acutely out of place at a conservative college in a very conservative part of the country, surrounded daily by students denying the existence of Covid, refusing to wear masks, carrying guns because they can. And while Charlie, Paul, and Antonia, all college professors, are living in more congenial parts of the country, as singletons, they've had to endure the past year profoundly alone. Meanwhile, those quarantined with partners and kids had the opposite challenge, of never being alone enough; the past year has been an intensive course in patience, compassion, and gratitude for kindness in small, shared spaces.

The pandemic completely upended all routines, all the accustomed ways of getting through each day. We came to see what a luxury reading can be—costing time and, especially, energy, resources that were in limited supply. What we wanted to read, what we *had* to read, whether we *could* read at all—depended heavily on circumstances. For working parents, reading for pleasure was often inseparable from reading with their kids. "I found myself reading more children's and young adult literature, because I was reading in tandem with my daughter," says Michelle, a translator and novelist. Laura, also novelist and college professor, has been reading Tolkien—60 min a day, out loud, in French—with her twelve-year old son. They started with *The Hobbit*, and are now nearly finished with *The Lord of the Rings*—a volume of fantasy reading she might not have undertaken on her own, but came more easily as the nurturing and educational multi-tasking of the home-schooling mom.

Meanwhile, "home-schooling" assumed nightmarish proportions for those of us in higher education. All plans for reading (never mind writing) were put on hold, as

---

[1] For anonymity, some names have been changed at the request of interviewees.

we scrambled to adapt to the technology of "emergency teaching." As Charlie explains, "in March, we were given mere days to 'pivot' to a 'fully online modality,'" bureaucratic jargon which translated to hours of reading about online pedagogy, hours of training from our instructional designers, hours struggling to master this platform, that app, this or that means of hybrid, hyflex, asynchronous, or synchronous content delivery. At the most frenzied moments in the past academic year, we could only do what Antonia calls "subsistence reading—the bare minimum required to teach." Most of our reading time has been taken up by student writing in the form of the usual essays, plus online asynchronous discussion forums, plus all the drafts, queries, and crises that seem to come in constantly via email and productivity chat-clients like Teams or Slack, meant to make us more approachable and accessible, functionally rendering us "on" for every waking hour. College faculty around the country are burnt out, with many reportedly considering leaving the profession entirely at the first opportunity [6]. Quarantine life didn't offer the life of the mind that many of us academics expected.

Reading, or any intellectual pursuit at all, took resources that we often just didn't have. Living through the pandemic unquestionably took a toll on everyone's wellness, and though we all somehow avoided Covid, we had an array of other health problems: anxiety and depression; tooth grinding, and weird autoimmune skin conditions exacerbated by stress; a chipped tooth; dizzy spells; kidney stones; sinus infections; suspicious moles. Not feeling well, and worrying about *why,* sapped everyone's energy and concentration. Others, already burdened with health problems of their own, were pushed to their limits when Covid struck friends or family. Adrienne, a technology worker and writer, had been struggling with health issues and anxiety for several years; then, she nearly lost her brother to Covid in April. Stress took a physical and mental toll. "Focussing isn't easy when you obsessively worry, and the pandemic breeds worry. It took months for my central nervous system to relax, months for reading to be pleasurable again."

Everyone was worried, unfocused—and our stress was compounded by the struggle to somehow feel differently than circumstances demanded. We kept apologizing to one another for losing track of time, for falling behind on work emails and social connections; we kept chastising ourselves for not being the workers, or readers, we thought we should be. More precisely; *on top of* looking after partners, kids, and ourselves, *on top of* keeping up with jobs which often seemed bizarrely abstract and irrelevant, *on top of* coping with our respective little corners of a terrifying global health crisis—we were convinced that we should be getting *more* done.

This pressure (always self-imposed) to read more, if not "better," afflicted all of us to some degree. Sarah was embarrassed that her reading wouldn't compare well with that of the imaginary "high achievers." As Mneesha, a college professor, says, "I perpetually feel guilty pressure from myself, that I should read more academic work and less for pleasure." Frank adds, "My failure to use this pandemic year to enrich myself, or at least entertain myself though pleasure reading, made me feel guilty, lazy, and intellectually incurious." And Antonia told me, "I have a stack of books for pleasure that has been accumulating since last summer that I haven't been able to read because my brain is simply tired from all the staring at screens, from all the painstaking but ultimately tedious and enervating work of grading student papers and replying to work emails, and from generally keeping the side up with political campaign work. Not being able to get to the pleasure reading makes me frustrated and sad…but at the same time I feel as though we are living somewhat out of time, and those books will be there when we enter time again."

Like Antonia, everyone reported being haunted by The Stack of Unread Books. When Sunny and her partner Matthew joined a zoom chat with me, they had each come prepared with their respective piles, pushing them toward towards the camera, like the material manifestation of literary sins, needing absolution. They were in abundant, penitent company: everyone had such a pile crouching reproachfully in some corner of the house. For many of us, if we could manage the flow of information that we had to read and write as part of our working days, *on top of* the intense onslaught of the news, there was little energy left over for sustained immersion in complex literary texts.

Often, the healthiest, most necessary, choice was to avoid reading altogether. Peter worked his way through several years' worth of the New York Times "mini" crossword, along with "mathdoku" puzzles, doing several a day—anything to avoid the news at its most intense; Sunny, similarly, found herself hooked on online Scrabble or similar word games "for easier, quicker, gratification" which she knew wasn't teaching her much, but which "felt smart".

Others craved stories, and when we didn't have the bandwidth to read them, we watched, consuming hours of television—documentaries, old movies, whatever new series was trending on.

Netflix, anything set in another country or language, allowing us the pleasure of vicarious travel from the comfortable prisons of our sofas. And we listened—avidly attentive to the stories narrated in audiobooks, and, especially, podcasts. Unable to go the gym or dance class (or anything, anywhere), I started going for long walks in Boston's Emerald Necklace Park network—trees, plants, water, and fresh air were revitalizing; podcasts kept my anxious mind from tying itself in knots. And although I wasn't reading per se, I could feel the influence of good writing in my favourite series, from the highly researched, carefully assembled documentary storytelling of This American Life and RadioLab, to the callers on Dan Savage's sex and relation-ship program, The Savage Lovecast. I wish that my undergraduate students would take as much care with their research papers as Savage's callers did—you could tell that people had turned their relationship questions into little essays, carefully com-posed, organized, and rehearsed. These podcast stories were not *Literature*, but were compelling in their authenticity and immediacy, nonetheless. When our social worlds had contracted into the confines of our little city apartments, with so little variety, so little chance for even the most mundane interaction with friends, cowork-ers, neighbours, deprived even of the pleasures of gossip [7, 8], the struggles and misadventures of these strangers offered a chance to feel, however imaginatively and vicariously, involved in the lives of others.

And even as we continue to explain away all the reading we meant to do and didn't…we read. In fact, one friend after another spoke apologetically about not reading enough/more/better…and then, in almost the same breath, characterized much of their reading in the past year as obsessive, even desperate. The problem was not that we weren't reading enough, it was a matter of (perceived) imbalance. Not enough reading for pleasure, not enough pleasure in the world—so we read for fun when we could, and we couldn't *not* read news, analysis, and commentary, the need to escape being (for many of us) insufficient justification to refuse to engage.

Reading about current issues wasn't exactly restful—indeed was more often painful and agitating—but it was also necessary, giving us the means to critically respond to the upheaval of the past year. The health disparities laid bare by the pandemic were very obviously bound up with all manner of injustice that all seemed to come to a boiling point in the past year—the climate catastrophe, income inequality, racism. Trapped in our homes (and arguably constrained by our own privilege), reading, at least, was one way to feel we were doing something: educating ourselves for future struggle if nothing else, and—for those of us in higher education—becoming better prepared to teach others.

We all share a deep faith in learning as the key to positive change. My Feminist Book Club has always focused its reading on questions of social justice; we had read *White Fragility* in 2019, and added *The New Jim Crow*, the documentary *Thirteen*, and *How to be An Anti-Racist* in the past year—a reading selection that is echoed throughout my community of readers. Charlie tackled works on class, inequality, and exploitation, as did Paul: from *Down and Out in Paris and London*, to *The Shock Doctrine*, *White Trash*, and *Hillbilly Elegy*. We also read about climate change, at least as much as we could manage—*Walden Warming*, *The Overstory*, *How to be a Good Creature*, *The Democracy of Suffering*. Liz describes these works as disturbing and stressful, but also "offering hope, and reinforcing the relevance of the citizen science work" that she does in her community.

Anne puts it best:

> Reading is moral to me, as well as being a sort of armor against a predatory culture: it's good to understand what they are doing to you, who is doing it, and why. And there is no more predatory, exploitative culture than the United States of America at this point in the empire. Watching my communities, my world, become an impoverished apocalyptic hellscape overnight, I feel obligated to know and understand who is responsible for this injustice.

This need to understand was energizing, educational, humbling…and—necessarily—challenging, underscoring the uncomfortable tension in our lives between wanting to keep ourselves and our loved ones safe, feeling called upon to participate in change, and yet feeling a variety of limitations about what to do, or how. Antonia found that the reading she did to keep up with the news sites, even at its most stressful, was nevertheless an essential way to create purpose in her day-to-day life. Antonia, like me, like most of my friends, is someone whose life revolves around texts of one kind or another; like me, a childhood habit of reading for pleasure led her into an academic life as an academic, researching and teaching at a large US university. Her "normal" life involves a lot of reading for diversion on top of what she reads for her teaching and research. But this past year was far from normal. The US election, in particular, galvanized her: she became a passionate follower of political news and analysis, and after years of avoiding social media, found that she loved Twitter—the immediacy of it, and its "dialogic aspects" allowed her to connect with people and information that she would otherwise have missed. Many of us find Twitter a cacophonous slough; several friends mentioned that they had to abandon Twitter for their sanity because the frenzied tide of overwrought commenting (along with a tedious trend toward productivity-bragging) was just too much. But Antonia discovered one of its strengths, creating communities around identity and intellectual interests (for her, #TwitterHistorians and #LegalHistoryTwitter).Here in the US, the pandemic seemed almost as though it were an unsurprising culmination of the past several years of the Trump administration's actively destructive policies, and as we moved toward the election, we found the coverage of callousness, incompetence, and paranoia gruelling. Some, like Peter, tried (but failed) to avoid the news entirely; others, like Adrienne, Laura and Liz, rationed their exposure—no tv, no radio, no social media, and only an hour or so a day allotted for reading coverage in respected news sites. Still others developed what was, by their own admission, an unhealthy compulsion. Frank, cut adrift in the midst of a major life change, and—as a biologist acutely aware of just how dangerous Covid could be—determined to leave the house as seldom as possible, found himself swamped by news coverage. "I became obsessed with the plethora of disasters we were forced to confront: a once-in-a-century pandemic; a failed government headed by an amoral, narcissistic, incompetent and malevolent president enabled by a complicit congress and an army of white supremacist, conspiracy-theorist know-nothings; all accompanied by a widespread unrest as the country reckons with its racist history and systemic biases." As so many of us experienced, the rage, frustration, and anxiety brought on by the world around us made sustained reading for pleasure often impossible; the reading that seemed to promise a way to feel connected, and somewhat in control of our circumstances, came at a high price in terms of peace of mind.

But still…we kept reading.

Because despite all the words we had to read—to meet work obligations, to stay informed, to feel engaged even as we also felt very, very trapped—we still needed to read for relief and escape.

Importantly, when it came to the reading we did for comfort, for fun, it mattered that we could follow our own tastes, desires, needs; that we could *choose*.

Many, unapologetically, choose things that were "easier" than what they would usually tackle. Adrienne ploughed through the latest book in the *Twilight* series: "Honestly, is it 'trash'? Is it cloying in some places? I don't know. I think people are too hard on pop lit. Was it easy to get lost in? Absolutely. In that moment, that's what I needed." For Anne, similarly, reading was vital for providing calming distraction: she tore through several novels of the "horror demon possession type, because, as sick as it sounds, I find horror stories to be much less terrifying than the real world." In contrast, Liz offset her reading of the news with a turn to comforting and familiar novels, like the *Harry Potter* series, "because I needed to read about people treating each other kindly in trying times."

Many of us found ourselves reading books we'd read (sometimes many times) before. As Laura puts it, "when it seemed like everything in my life had changed, reading and re-reading became a consolation in more ways than one, as proof that something in my life could remain *un*changed." So, we looked to favourites from our teenage years or younger—*Harry Potter*, of course, as well as *Nancy Drew*, or *A Wrinkle in Time, Winnie the Pooh*. We re-read favourite "genre" fiction—detective novels, suspense and espionage, historical romances. We brushed up on Molière. We had to re-read *The Plague*. We explored short fiction and poetry in literary magazines. We tackled big series, or foreign language novels, or, in the case of Elena Ferrante's series, both at once. Even when the story was new, our ability to read and enjoy was a reassuring constant.

Time during the pandemic became strange—one day bleeding into the next, feeling individual minutes weighing on us and yet somehow never having enough hours in the day. And yet—we found the time for escape. Antonia, out for a walk, found a copy of *The Hunt for Red October* in a Little Free Library, and is now re-reading Tom Clancy's *Jack Ryan* series, books she had enjoyed as a teenager, but which she wouldn't "normally" allow herself now. Instead, "I find they delight me. With my adult and professional eye, I have a new appreciation for the amount of meticulous research they required, a surprisingly progressive approach to gender, elaborate plots, and I am now amused by how quaint all the technology seems today, and how much harder spycraft was before the digital age." Kaycie, unable to work much because of the pandemic, revived her childhood love of science fiction and fantasy. Like Antonia, she found time to immerse herself in the lore of each story in a way she wouldn't have been able to in the midst of managing her escort business. Now, "I was able to take all the time and mental energy needed to fully re-acquaint myself with the characters, understand all the political manoeuvring," and with help of fan wikis appreciate the elaborate world-building in series like *The Wheel of Time* or *The Bobiverse*.

Everyone knew that these choices were motivated by a desire, a need, to escape, to be in another place, another time. As Antonia says, "Ever since childhood, I knew that as long as I had a book, there would always be someplace I could *go*. Pandemic times have only amplified that feeling." Even if the book is sitting in its pile, waiting, "it's there if I need to *go* somewhere." In a year when we've spent more time in our homes than we ever thought possible, when the only remarkable thing about a given day was how identical it was to the one before, we were all starved for stimulus and change. Like our choice of TV shows and films, stories in other settings, real or imagined, gave us the illusion of being somewhere new, of being able to travel, to *go*.

But immersing ourselves in different worlds through our reading wasn't just diversion and escape -crucially, it was also a way to experience hope. As Kaycie put it, citing a recent article that we had all read [9]: "fantasizing about future fun, when we'll be able to travel or gather in groups again, is good for our mental health. Fantasy planning stimulates the same parts of the brain that light up when we're actually having in-person experiences, and we can get a surge of the same feel-good chemicals that come from being with others or doing things that bring us joy."

Being with others. Joy. Hope. The real world brought us little enough of that in this past year, our Year of Covid. When so much was lost, so much was out of reach, what got us through was our connection to one another, both in person (in such rare and limited ways) and through our imaginations. Reading, as my friends all insisted, was vital in feeling connected, even as we were (still are) unable to be together.

Paul, living alone through the pandemic, says, simply, "My books are my friends." And Kaycie says, "In many ways, my books have been a replacement for social connection, care, love, and adventure that I'm currently missing in my life. Amid considerable isolation, reading, getting to know the characters, has given me the experience of getting to know a friend intimately, and at times feeling the emotional connection of a lover. It's been a balm during difficult times."

As Anne put it, "I say, without exaggeration, that being able to read through this past, horrible year likely saved my life."

References

1. Gilbert DT. Stumbling on happiness. New York, NY: Vintage Canada; 2006.
2. Beck J. We have to grieve our last good days [Internet]. The Atlantic. Atlantic Media Company; 2021 [cited 2021 July 2]. Available from https://www.theatlantic.com/family/archive/2021/03/we-have-grieve-our-last-good-days/618233/Atlantic.
3. Odell J. How to do nothing: resisting the attention economy. New York, NY: Melville House; 2019.
4. Kwon RO. Trouble focusing? Not sleeping? You may be grieving [Internet]. The New York Times. The New York Times; 2020 [cited 2021 July 2]. Available from: https://www.nytimes.com/2020/04/09/opinion/sunday/coronavirus-grief-mental-health.html.
5. Stern J. This is not a normal mental-health disaster [Internet]. The Atlantic. Atlantic Media Company; 2020 [cited 2021 July 2]. Available from: https://www.theatlantic.com/health/archive/2020/07/coronavirus-special-mental-health-disaster/613510/.
6. Fidelity investments and the chronicle of higher education study: more than half of college and university faculty considering leaving teaching, citing burnout caused by pandemic [Internet]. Business Wire; 2021 [cited 2021July 2]. Available from: https://tinyurl.com/ypcnb9su.
7. Sullivan L. 50 Shades of Shade. This American Life Episode 730 ("The Empty Chair"). [22 January 2021]. Available from: https://www.thisamericanlife.org/730/the-empty-chair.
8. Rosenbloom S. Travel and the art of anticipation. New York Times. [2021 5 February]. Available from: https://www.nytimes.com/2021/02/05/travel/future-trips-anticipation.html.
9. Rosenbloom S. Travel and the art of anticipation [Internet]. The New York Times. The New York Times; 2021 [cited 2021 July 2]. Available from: https://www.nytimes.com/2021/02/05/travel/future-trips-anticipation.html.

Without Words: The Art and Therapy of Grief and Loss in Pandemic Times

5

Stephen Legari

> *Ring the bells that still can ring*
> *Forget your perfect offering*
> *There is a crack, a crack in everything*
> *That's how the light gets in*
>
> —*Leonard Cohen, Anthem*

Situating the Author and Chapter

The COVID-19 pandemic reached Canada and my home city of Montreal in full force in the early days of April 2020. We watched with gripped anticipation as our friends and colleagues in China and then in Europe suffered the anxiety of confusion, the loss of freedoms, and the loss of loved ones on a massive scale as the pandemic ravaged and raged. Governments both local and national took important measures in attempts to limit the spread of the virus which ultimately saved many lives. These measures however also seemed to halt the very breath of culture itself and render inaccessible those vital rituals which maintain a sense of meaning in life and community.

The delivery of care in the support of individuals, families, and groups was radically impacted by the pandemic. In the early days of spring 2020, therapists and counsellors of every modality scrambled to adjust to a rapidly accelerating new reality. The ubiquitous grid of the virtual meeting that we now take for granted was then still a new frontier for providing psycho-social support. This was no truer than

S. Legari (✉)
Division of Education and Wellness, Montreal Museum of Fine Arts, Montreal, QC, Canada
e-mail: slegari@mbamtl.org

© The Author(s), under exclusive license to Springer Nature Switzerland AG 2022
M. G. Marini, J. McFarland (eds.), *Health Humanities for Quality of Care in Times of COVID -19*, New Paradigms in Healthcare,
https://doi.org/10.1007/978-3-030-93359-3_5

in providing comfort to the bereaved. It has been suggested that we are collectively grieving the loss of the lives we knew before the pandemic, but for those who have had to endure the loss of a loved one as a direct consequence of the virus or not, have been confronted with unforeseeable obstacles to the grief process. These obstacles have included both literal barriers, such as not being able to physically access the dying or deceased, and barriers to ritual and culturally essential practices in meaning-making.

My humble ambition in writing this chapter is to capture a snapshot of how the fine arts, creative arts therapies, and community arts have helped to bridge a morass in the providing of palliative and bereavement care during a global pandemic. At an unprecedented time when virtual connection has supplanted the very real requirements of human connection, the arts have given us access to our shared humanity. By way of transparency and bias, I am making these claims while working as an art therapist in a major fine art museum who has been working virtually for more than 15 months at the time of writing. In the absence of real-life access to both art objects and art materials as well as the absence of real-life group participants, I have been left to wonder just how effective a virtual format of group therapy could be. More generally I have worried for the immense work ahead in individually and collectively processing such a complex grief as has been endured. What I believe I have learned over the past 15 months is that the arts are not simply a means of coping but in fact a way through.

My goal here is not one of providing scholarly evidence but as call to recollection and reconnection with the arts as a conduit of continuity, narrative, and compassion.

This chapter will explore the use of art in the expression of grief in the time of the Covid-19 pandemic. We will look at both the specialized use of art-by-art therapists in palliative care and bereavement and how the pandemic has changed their work. We will examine the use of the fine art object in virtual format as a vehicle for expressing the complexities of grief. And we will also look at the intuitive ways that artists have used craft, artmaking, and collective work during the pandemic as a means of both memorializing and bringing meaning to a phenomenon too large and too devastating to capture in words.

The interviews, artworks, news stories, and anecdotes in this chapter are all derived from the context in which my work takes place. And while these stories are relevant to a North American context, it is my hope that the themes will find a universal resonance. The examples cited for instance of artworks from my own museum's collection are intended to inspire the reader to look deeply into the collections of the museums both near and abroad. The practice of art therapy in the situations cited can be extended to those places where art therapy not only maintains presence within healthcare but more broadly to anywhere the arts and health have a history and present. The examples drawn upon to illustrate how people in community use art to process grief are echoes of what has been shared from every corner of the globe over the past year and more. The ringing of bells, the word that finds poem, the sharing and intimacy of music, and the rendering of a feeling through material into visual form know no borders in our collective human expression.

Art in Death and Dying

It is beyond the reach and intention of this chapter to report on the scope of research supporting the various uses of the arts in end-of-life care and bereavement care. The reader is referred to the most recent literature available from both practitioners and scholars [1, 2]. Likewise, the author acknowledges the imperative of approaching therapeutic work around grief with cultural humility and a broad appreciation for the multitude of ways that people the world over deal with death, but cannot provide here an anthropological review with the depth that this subject merits. Instead, a short review of the literature of the use of art in the grief process is provided to give the larger discussion context.

While the use of art, object, and ritual to express grief is as old as civilization, the examination of its role in clinical work is a more recent activity. Like their counterparts in bereavement counselling, art therapists have moved beyond the writings of Freud and Kübler-Ross to embrace more holistic, constructivist approaches to grief and palliative work. With sufficient social supports in place, including access to family, ritual, and community, most people will go through a grief process without the need for therapeutic intervention. However, complicated grief can have a damaging effect on the bio-psychosocial wellbeing of an individual, family, and community. Here, art-based interventions find fertile ground in a grief treatment process that centres on meaning-making, reconstruction of identity, and narrative [3, 4]. This chapter posits that the inaccessibility to cultural norms of death rituals imposed by the conditions of the pandemic, including but not limited to gatherings in significant spaces, not being able to be with the body of the deceased, the sharing of song and story, the sharing of food, the lending of care through touch, and other meaning-making activities, will lead to widespread complicated grief.

Art therapists use art-based interventions, non-directive art-making, and craft to enhance therapeutic alliance, safe-space creation, and the meaning-making potential for those at the end-of-life and the bereaved [4]. As Weiskittle et al. summarize, "Art therapists' orientation toward externalizing processes and facilitating insight meld naturally with meaning-focused therapeutic practice. The spontaneous creation of art, poetry, and performance offers a way to memorialize the relationship with the deceased and facilitate continuing bonds…theoretical models of art therapy posit that creativity is both a restorative and assertive act" [1].

Specific to the use of visual art, art therapists have explored and reported on the use of storyboard for re-construction of self-narrative [3], photographs [5], and a careful, progressive approach to providing structured to fluid art materials to clients [6].

Art therapists have also turned towards art materials themselves in heuristic explorations of their own grief process [7, 8] as a means of deepening their understanding of their clients' experience of grief and their own subjective experience [6], and in the regulation of their own active grief to better care for clients [9]. What appears to be universal is the assertion of those who use the arts in end-of-life and bereavement care is that they facilitate the externalisation of a highly complex experience where words may be, and often are, insufficient. And while there has yet to be literature produced on the use of the arts in facilitating the grief process during the COVID-19 pandemic, virtually or otherwise, the adaptation of practice using many of the aforementioned best practices has forged ahead [10].

Art Therapy in Palliative Care and Bereavement

The following are excerpts from two interviews with two art therapists working in palliative care centres. The interviews were conducted in March 2021, approximately 1 year since the pandemic arrived in earnest in Canada. Both art therapists work full-time in palliative centres in Montreal, Quebec, Canada. Both use visual and creative arts as a means of accompanying people in end-of-life and bereavement care. Both therapists have continued to work through the Covid-19 pandemic both on-site in the presence of their clients and families as well as virtually with the bereaved.

In their respective centres, they will work both bedside and in a dedicated art studio depending on the needs of the client and their family members. In the art studio, there is a myriad of materials available for clients to choose and explore with. The work may change considerably whether they are working with a child, with a bereaved spouse, with an adolescent, or with a group. The artmaking experience is believed to help the individual connect with something essential. Both employ arts-based legacy work that can take on various forms but may include memory boxes, photo collages, and the casting of hands of the dying patient often entwined with those of loved ones. In the two interviews, the themes of coping, choice, empowerment, pleasure, connection, communication, intuition, and creativity were predominant.

Interview with Sarah Tevyaw, Art's Therapist

What do you feel art contributes to supporting people in end-of-life care?

The first word that came to mind was choice. [When a patient arrives, I wonder] whose decision was it to come here? And how did you decide that as a family? Coming here is the last big decision you'll make, and you probably will not be leaving this place. And I think art brings an opportunity to start to explore choice a bit more. Even little things, like if a patient comes to the to the loom and chooses a colour of yarn [Sarah's art therapy studio includes a full-size loom for weaving that both patients and staff use and that collectively produces weaves that are hung in the studio.] *On its own, it was probably the biggest decision that they made throughout that day. [Play and pleasure were] actually the second words that came to mind. I always joke with patients, especially those that I get to know best, [asking them] "did you ever imagine that you would be coming here and doing this?…Painting or knitting or crocheting…? I don't think they ever thought that going into a palliative care residence and facing their end of life could be playful. It alters people's space you know. Just a couple of weeks ago [I went to see] an older lady who was pretty withdrawn. I met with her and proposed the idea of doing something together and we started painting. So every week, or even sometimes twice a week, I would go into her room and bring a small set of watercolours, a sketch pad, a paper cup, and two brushes, and she would paint, and we started filling her wall beside her with her paintings. Suddenly she changed her space around her by engaging in something so creative. It's time away from illness. I don't think we recognize how meaningful it can be for people just to do something completely other than talking about their sickness, their illness, or their approaching death. It's a wonderful escape from reality and I'm not sure I would have said that as a beginning art therapist, but now I recognize how absolutely important it is. It's just way too overwhelming to constantly sit with your illness all day.*

And [it's] a time to reminisce. Bringing art material to the bedside, accompanying people as they are entering their final weeks or final months or final days just allows the opportunity to talk, to not think too much and to not censor themselves too much. But just to remember and to talk about their life and their experiences and things that were rocky and things that went well, and things they may regret, or things they may want to repair. [And art also offers] a space to adjust to their new reality.

Can you tell me about the art studio and what happens there and how COVID has affected those activities?

I haven't yet been able to have open studio hours [when patients or family members can drop in and create]. Covid has made this place so quiet because it's so restricted. We become used to this stillness at the residence that never was there before, so I guess I'm even more hypersensitive to those particular sounds. I can listen to the sounds of peoples' feet shuffling or wheelchairs or walkers and I know that it's patient related and it's not just somebody who's kind of lost in the halls. Patients [will still] come and explore the studio. There's materials out on the counters there's materials accessible on the walls in crates ranging from pretty much anything you can imagine: natural materials, recyclables, crayons and strings and yarn. There's [also] the loom which has really been amazing to have.

As [patients] begin to decline, the artwork changes and especially with textiles I find people who are knitting, crocheting, sewing, embroidering…as they get weaker and maybe more confused or have more difficulty breathing, things unravel or become super knotted. If [patients] want to have an individual session, often times it's still at the bedside because they're in a mandatory isolation for two weeks.

Can you tell me about the differences or similarities of working as an art therapist in palliative care and bereavement?

For me it's almost like two different fields. To be accompanying someone until they die, I'm working with them quite closely and developing a relationship and knowing that it's going to end and then in a whole another realm I'm working with those who are left behind and who are missing those people. And I feel like I'm in a bit of a privileged state you know, to have both, because when I work with people in bereavement, I use all of my experiences of working with people who have died to help me.

How has the demand for bereavement support changed during the pandemic?

There's a growing need in the community. I've had people reaching out that had no connection to the [centre] and are looking for bereavement support for their kids and we've opened that possibility up now because we [now] have a space to do that. People are more isolated, so parents are less sure what to do. The kids that I've seen most recently maybe would have had more social support and their parents wouldn't have necessarily gone for a professional route because they would have been supported by their family or wouldn't have seen so many signs that kids are struggling. I think that lack of support that would have been so close to them is really evident right now. Their grandparents are not home with them and helping them and talking with them. Their experiences are far from normal these days and they don't have the support of their peers so I'm not sure if parents are recognizing that there is a greater need because those things are so absent. Kids [in bereavement] are just more anxious now...they open up about what it's like to not have their peers with them.

Interview with Karine Bouchard, Art Therapist

What does art bring to working with people in end-of-life care and bereavement?

I really see how it can touch people, how it can open doors to talk about emotions but also to create something like a tangible memory that we can keep...an object can imbue the presence of a loved one. I think that there is bereavement work that can already begin in palliative care, if only the mourning of the patient's autonomy, the mourning of the life before, the mourning of the house, the mourning of leaving all these things to come and settle here in a room. And with the family, it's also a mourning to see how the person has been transformed through the illness, how the relationship between the people has also been transformed through the illness - the impermanence of saying: this person won't be with us forever, so we can...prepare something together with a family, for example at the patient's bedside, [such as making] a box of memories together, then [ask] are there any symbolic objects or images that are important to you? The person who is going to die was present at the creation of the legacy object, so that's one of the avenues in palliative care or in bereavement that I see as being very important...it can be a box of memories, it can be [hand] casts, it can be a collage with photos of the family or with images that represent their relationship.

There's also the whole more exploratory part. Sometimes people just need to get into the creative process...the pleasure of putting paint on a big canvas together...the sensory side, taking clay in your hands. [There are people who say] "I've never done that I'd like to try it I'd like to try watercolour before I die." [We can] explore emotions with different artistic mediums, go into the unknown and see what emerges. As the art therapist, I am there to witness and to really validate them in what emerges and to normalize—that it's normal that you feel this way right now in relation to your mourning.

What has changed in your work during the pandemic?

When we enter a room of someone who arrives from the hospital, who is in isolation 14 days and we're wearing our astronaut costumes [COVID protocol Personal Protection Equipment] *...it's a challenge. To build a bond of trust, it's more challenging. There's a nurse here who took photos so that we could show what our face looks like because with the glasses and masks, the gloves, it's certainly a bit intimidating. In art therapy we work with all manner of artistic mediums, so it's also a challenge for disinfection and to always disinfect between each session, which adds a weight on the daily work.*

People used to gather here [before the pandemic] at the day centre. It was a place to break the isolation for people who have been diagnosed with an incurable disease. At the moment it's the biggest and most difficult thing I see, is isolation, isolation, isolation. Then of course there's all the regulations from public health which means there are also restrictions on visitors...

A nurse and I made a hug wall made of plastic [so that people could hug each other safely]. It's better than nothing but you know it's like all these things, we have to disinfect the hug and there's no spontaneity...

The virtual [bereavement] work doesn't allow us to offer all the materials we would normally have. At the same time there are some participants are using materials we wouldn't normally be able to offer like using personal photos or Photoshop...it opens up other possibilities. And they are making art in the privacy of their home space, and they may return to work on the artwork during the week between art therapy sessions which encourages a continuity in their mourning process.

What about your own creative work?

I would say that in the last year we've been in a bit of emergency mode here in the health care system and I've been called upon to wear several hats. But I am working on an illustration project. It's for a book of poems that will be in the patients' room that people will be able to keep after their loved ones' stay in palliative care, and also as a kind of transitional object that they will be able to look at with the person at the end of their life.

I don't want to speak on behalf of all the therapists, but at the moment, with the crisis, it's a challenge to take time for my creative work. I used to do concerts at the little bar on the corner of my street, but now it's closed. When the time comes, I trust the creative process that will emerge naturally, once again.

Grief and the Fine Arts Museum During the Pandemic

In the wake of public art created in New York City in response to September 11, 2001, James B Gardner reflected on the role and responsibilities of museums following a collective experience of trauma, crisis, and tragedy. He posed the following critical questions:

> What role should museums play in a time of crisis? What public expectations would we face? What new responsibilities would we have? What role should we play in constructing

collective memory? And, perhaps most importantly, how would we address our responsibilities as museums within that tragic context, at the difficult intersection of grief and history? [11]

Museums, and every manner of cultural venue, closed in the earliest days of the pandemic in Canada. These measures mirrored those from other parts of the world where the pandemic had already taken hold. The impact of these closures is yet to be fully appreciated but we suggest that the cutting of access to tangible cultural experience has compounded the pandemic grief experience. Art, in addition to being a mirror that helps reflect the world back to us, is also a vehicle to making meaning in a senseless world. What research has revealed is that museums, along with opera houses and theatres, are ideally suited to supporting visitors with a very low risk of transmission of COVID-2, lower than any other indoor activity as it turns out, as long as the well-known sanitary precautions are in place [12]. Whereas many commercial spaces were experimented with using a revolving door approach to safety and priority, museums on the whole were forced to remain closed. There is a broader discussion to have concerning the contribution of culture to an economy, but this will have to wait for the pandemic's post-mortem. What we propose is that the reality of museum's closures over the past 12 months has negatively contributed to a collective experience of grief and the missed opportunities to mitigate its severity due to the lack of access to art, space, light, contemplation, and (safe) social connection.

Since the first lockdowns and closures, museums across the world have mobilized virtual content to fill the void left by the inability to welcome visitors and participants in person. This past year has seen a myriad of offerings from immersive virtual visits to online pedagogical resources, to critical discussions on the future of museums presented in the all-too-familiar grid of online interaction. Those museum programs dedicated to providing access to wellness-based activities and art therapy also adapted to the new online reality. What made these offerings possible was the digital collections; those thousands of hi-resolution images of every manner of artwork and object that, despite the virtuality of access, still seemed to hold a potential to help people connect, project, and share their experience.

Exploring the fine arts through the pandemic has provided a forum for people to mourn what is missing and who is missing. The collections have provided respite, escape and opportunities to grieve. Grief has long been an important theme to explore for many people who have dealt with illness, disability, social exclusion, and trauma and the fine arts can serve as a fulcrum for these difficult discussions. This is perhaps nowhere truer than in actual bereavement work that incorporates the fine arts in its model.

At the Montreal Museum of Fine Arts, we have developed an art-therapy bereavement model for small groups in collaboration with community and end-of-life care organizations. This model has progressively moved away from a staged or task-oriented framework to provide an arts-based forum in which each participant can explore their grief at their own pace. This model includes exploring images from the museum's collection, creative activity, reflection, and facilitation by art therapists and other helping professionals. Museum-based art therapy has been gaining momentum in the last decade but the literature has not elaborated specifically on the use of its collections in bereavement work [13–18].

Recently, we concluded a pilot project virtual version of this model in response to the pandemic conditions. A 10-week program was conceived in collaboration with a palliative care centre (see interview with Karine Bouchard) using a popular online meeting software. As with our on site model, the bereavement sessions included viewing different artworks from the museum's collection, creative responses, and a closing reflection. Each weekly session was built around a theme relevant to the bereavement process, but flexible enough to be interpreted as needed by each participant. Themes included: navigation, the path, relation to the deceased, the body, identity, and the journey. We found it was not necessary to press participants into the difficult work of sharing their stories. The artworks, both those from the museum's collection and those created in the intimacy of the participants' homes where they joined us from virtually was a sufficient way in and sometimes way out of the intensity of the bereavement process.

There seems to be no limit to the kinds of artwork or periods in art history that can be used in eliciting the grief narrative. While care is taken in selecting the works to be viewed, the inherent subjectivity involved in relating to artwork is mitigated by innumerable factors including the subjective states of the participants on a given day, the comments of other participants, tastes in art, and in this context, the overarching impact of the pandemic on people's capacity to maintain their own wellbeing. The following two examples are but a drop in the ocean of what has been shared over the past year, but they have been used more than once in discussing grief and its journey. The natural world seems to hold particular value in these discussions and in the elicitation of personal anecdotes of connection. By contrast, the unforgiving power of nature can also yield meaningful reflection and inform selfawareness and re-consideration of self in the face of its vastness.

The theme of navigation (Fig. 5.1) can be a useful way into discussing the grief process. It at once validates the inevitable waves of grief that seem to arrive and abate without warning and gives the participant permission to simply cope when needed. van de Velde, *Kaag and weyschuit, fleeing the storm* can be a near overwhelming depiction of unrelenting waves that seem to threaten the small and vulnerable looking crafts and their sailors. Upon closer inspection, group members often note the contrast between the raging sea and the light emerging from behind the clouds in the upper left of the canvas. The appearance of faint ships in the distance can invite discussions of perseverance and the realization that the sailors are navigating out toward the storm rather than retreating from it.

The path (Fig. 5.2) is another important theme that invites group members to situate themselves in the scene depicted. There are countless appearances in art of roads, trails, routes, and lines that can help in discussion and can provide a bridge to the larger theme of the grief journey. Here Cézanne's *Bend in a Road in Provence*

Fig. 5.1 Willem van de
Velde, *Kaag and
weyschuit, fleeing the
storm*, Around 1685, Oil
on canvas, 41.2 × 54.5 cm

Fig. 5.2 Paul Cézanne,
*Bend in a Road in
Provence*, Around 1866 or
later, Oil on canvas,
92.4 × 72.5 cm

gives group members the opportunity to project onto or into the represented path. Sometimes there are unnerved reflections on the uncertainty of what lays ahead. Other participants have reflected on the strong forms flanking the path including the rock and the tree. These forms may invoke the deceased. What is almost universally shared is the sense that the road ahead is long but perhaps manageable.

The Art of the Memorial

Memorial, from the Latin *memoriale* 'record, memory, monument.' The Covid-19 pandemic is very much a reality in the better part of the world at the time of writing. Despite the rollout of vaccines in the most affluent nations, the sustained state of restrictions and threat remain the norm. There is a human need to find a way to grieve together when a catastrophe takes many lives. War memorials, museums dedicated to the history of atrocities, monuments to lost souls, and the commemorative object are found throughout the world and the therapeutic use of their potential has been mobilized to the benefit of individuals and collectives included [7, 19, 20]. To the global tragedy before us, overtures to Covid-19 related grief have begun to appear in form of flags at half-mast on public buildings and the more arduous, and essential work of honouring the dead by name [21].

And while the more official memorials to come will certainly adorn different institutional and public spaces, both professional artists and those dedicated to community arts have already begun the difficult work of meaning-making with and among their communities. The following is a very limited sample of those stories that have reached our attention by means of the press, but by no means should be considered comprehensive. Their inclusion pays tribute to how grief is honoured both in art and craft and, despite the inability to gather in person, helps us to connect to a collective sense of shared pathos. These stories will invariably fade from news sites and even their URLs may go quiet. Such is often the nature of community-based bereavement art. Its presence may be ephemeral, but its impact is enduring. It weaves threads of connection often without the need for funding, sanction, or attention. It speaks to a more ancestral and intuitive mode of expression and collectivity that provides solace and meaning.

Contemporary artists have used both technology and traditional materials to address a critical and collective sense of loss. Lozano-Hemmer, a Mexico City-born, Montreal-based artist created an AI-operated machine that transforms photographs of people who have died from Covid-19. The public submits the photo through his website and the machine plots out the portrait in grains of sand. In a true poetry, the portrait is only temporary, and the sands pass through the hourglass. The entire process is livestreamed [22]. Eric Waugh, also a Montreal artist, completed a painting of 200,000 hearts on a massive 10-metre-wide canvas in tribute to those who have died of Covid-19 in the US [23]. Robin Bell of Washington, DC sent a beam of light out of his apartment window. Projected onto the brick wall of a shop across the street appeared the words "Covid Memorial." Below them scrolled a slideshow of faces of covid-19 victims, along with messages from their loved ones [24].

No discussion of grief and the arts would be complete without citing the remarkable capacity of community members to instinctually create together to create a network of care and support in difficult times. During the pandemic there has been an almost constant call to gather, virtually, around the activity of making. The ancestral practice of crafts and their inherent qualities of transmission and connection have not only found a resurgence in the hobby hands of new knitters, quilters, carvers, and embroiderers but these skills have been turned to the difficult work of processing the often-absurd sense of grief that has permeated the pandemic.

Montreal seniors have knit colourful scarves to commemorate the lives of the 251 long-term care residents who lost their lives during the first wave of covid-19 [25]. Three women in Ontario, Canada began a project to knit a memorial that will be 836 square metres when finished and knitters across the country have volunteered to help with this expression of community grieving [26]. This is reminiscent of the AIDS Memorial Quilt, created in 1987, the largest piece of community folk art in the world as of 2020. In Los Angeles, a gallerist invited the public to contribute to the Memorial Crane Project, adding an origami crane for each life lost, along with personal stories [27]. The New York Times invited people of every age to share a photograph and story of someone they lost in the preceding year of the pandemic in a project called "What Loss Looks Like" [28].

Discussion

Given the opportunity, humans will gravitate, intuitively, towards the creation of meaning in the face of tragedy. One of the more vexing circumstances of Covid-19 has been the condition of isolation in the face of grief and loss. As the pandemic has stretched well beyond a year in every part of the world, the compound experience of grief in the face of the loss of autonomy and life has created a complexity of condition that will take generations to resolve.

This chapter has posited that art is ideally suited to navigate the complexities of bereavement. Art gives form where there is only an amorphous sense of absence. Art holds a space where words are not yet ready to arrive. Art captures the immensity and the personal concurrently. In discussions with art therapists working through the pandemic we learned that art, whether analogue or virtual, was the medium through which the work of supporting people through end-of-life and beyond remained imaginable. Through the virtual use of the fine art collections of museums, we learned that people may still gather around a shared experience of wonderment and express their grief within a held frame of support and shared meaning. And through contemporary and community-based art we were reminded of the innovation and ingenuity of art makers to create work that challenges isolation and despair to create collective records of memory in thread, light, and megabits.

References

1. Weiskittle RE, Gramling SE. The therapeutic effectiveness of using visual art modalities with the bereaved: a systematic review. Psychol Res Behav Manag [Internet]. 2018;1(11):9–24. Available from: https://pubmed.ncbi.nlm.nih.gov/29440940.
2. Wood MJM, Jacobson B, Cridford H. The international handbook of art therapy in palliative and bereavement care LK—https://mcgill.on.worldcat.org/oclc/1084618690 [Internet]. NV-1 online resource (xxviii, 421 p). New York, NY: Routledge; 2019.

3. Lister S, Pushkar D, Connolly K. Current bereavement theory: implications for art therapy practice. Arts Psychother [Internet]. 2008;35(4):245–50. Available from: https://www.science-direct.com/science/article/pii/S0197455608000610.
4. Beaumont SL. Art therapy for complicated grief: a focus on meaning-making approaches. Can Art Ther Assoc J. 2013;26(2):1–7.
5. Weiser J. PhotoTherapy techniques in counselling and therapy -- using ordinary snapshots and photo-interactions to help clients heal their lives. Can Art Ther Assoc J. 2015;15(17):23–53.
6. Garti D, Bat Or M. Subjective experience of art therapists in the treatment of bereaved clients. Art Ther [Internet]. 2019;36(2):68–76. https://doi.org/10.1080/07421656.2019.1609329.
7. Kalaba E. Making memory and meaning: the memorial fuction of art [Internet]. 2009. (Research Paper). Available from: https://spectrum.library.concordia.ca/976546/.
8. Thompson C (Christy). Honouring loss: using mementos as transitional objects to explore ambiguous loss [Internet]. 2019. Available from: https://spectrum.library.concordia.ca/985636/.
9. Arnold R. Navigating loss through creativity: influences of bereavement on creativity and professional practice in art therapy. Art Ther. 2020;37(1):6–15.
10. Carr SMD. Art therapy and COVID-19: supporting ourselves to support others. Int J Art Ther Inscape [Internet]. 2020;25(2):49–51. https://doi.org/10.1080/17454832.2020.1768752.
11. Gardner JB. September 11: Museums, spontaneous memorials, and history. In: Margry PJ, Peter J, Sánchez Carretero CTA, editors. Grassroots memorials: the politics of memorializing traumatic death [Internet]. NV-1 onl. New York: Berghahn Books; 2011. p. 285–303. (Remapping cultural history; 12). Available from: http://site.ebrary.com/id/10497646.
12. Kriegel M, Hartmann A. Covid-19 contagion via aerosol particles – comparative evaluation of indoor environments with respect to situational R-value [Internet]. 2021. Available from: https://doi.org/10.14279/depositonce-11401.2.
13. Legari S, Lajeunesse M, Giroux L. The Caring Museum/Le Musée qui soigne. In: Jury H, Coles A, editors. Art therapy in museums and galleries: reframing practice. 1st ed. London: Jessica Kingsley Pub; 2020. p. 157–80.
14. Canas E. Cultural institutions and community outreach: what can art therapy do? Can Art Ther Assoc. 2011;24:30 3.
15. Ioannides E. Museums as therapeutic environments and the contribution of art therapy. Mus Int. 2016;68:98–109.
16. Treadon CB. Bringing art therapy into museums. In: The Wiley handbook of art therapy; 2015. p. 487–97.
17. Treadon CB, Rosal M, Wylder VDT. Opening the doors of art museums for therapeutic processes. Arts Psychother. 2006;33:288–301.
18. Salom A. Reinventing the setting: art therapy in museums. Arts Psychother. 2011;38:81–5.
19. Betts D, Potash J, Luke J, Kelso M. An art therapy study of visitor reactions to the United States Holocaust Memorial Museum. Museum Manag Curatorsh. 2015;30. https://doi.org/10.1080/09647775.2015.1008388.
20. Cowan B, Laird RA, McKeown JTA-TT. Museum objects, health and healing: the relationship between exhibitions and wellness LK - https://mcgill.on.worldcat.org/oclc/1113229629 [Internet]. NV-1 online resource. London: Routledge, Taylor & Francis Group; 2020. (Routledge research in museum studies). Available from: https://www.taylorfrancis.com/books/9780429467813.
21. https://www.nytimes.com/interactive/2020/obituaries/people-died-coronavirus-obituaries.html.
22. https://memorialcovid-lozano-hemmer.web.app/#!.
23. https://montrealgazette.com/opinion/columnists/brownstein-montreal-artist-shows-a-lot-of-heart-in-covid-19-memorial-tribute.

24. https://www.npr.org/2020/05/26/861886215/memorializing-those-who-died-in-the-time-of-covid-19.
25. https://www.cbc.ca/news/canada/montreal/collective-grieving-through-knitting-1.5944643.
26. https://www.cbc.ca/news/canada/toronto/three-knitters-project-covid-19-memorial-blanket-canada-art-installation-1.5663792.
27. https://memorialcraneproject.org/mission.
28. https://www.nytimes.com/2021/04/06/insider/covid-grief-loss.html.

Spirituality as the Basis and Foundation of the Medical Profession

6

David Cerdio Domínguez

> *"Where there is love for medicine, there is love for humanity"*
>
> *Hippocrates [1]*

Throughout history, humanity has faced many challenges, however, never before have we all found ourselves united against the same adversary [2]. SARS CoV2 pandemic has revealed both the greatest weaknesses and the greatest strengths of human beings; on the one hand, we have experienced disinterest, fragility, fear and social selfishness, while on the other hand, there have been glimpses of the very best of man; the love shown by the vocational work of our health professionals and their personal sacrifices. Perhaps that is why there is no better time to reflect on the profession that Edmund D. Pellegrino called "the most humane of the arts, the most artistic of sciences and the most scientific of the humanities" [3].

What is a Physician?

A physician is any man or woman who, eager for science, wishes to grasp the tragedy in human destiny [4]. The doctor is driven by the desire to put all their talents at the service of mankind, and recognizes that there is no greater honour than being able to serve a sick person in their moment of maximum vulnerability; paradoxically our fragility leads to our greatest strengths. A doctor accompanies, comforts,

D. C. Domínguez (✉)
Faculty of Health Sciences, Universidad Anáhuac México, Huixquilucan, Mexico
e-mail: david.cerdio@anahuac.mx

M. G. Marini, J. McFarland (eds.), *Health Humanities for Quality of Care in Times of COVID -19*, New Paradigms in Healthcare,
https://doi.org/10.1007/978-3-030-93359-3_6

and accepts that alleviating pain and suffering, with all its implications, are an essential and fundamental part of his/her way of life.

A doctor lives a intensified life, with glimpses of both human happiness and life but also human misery, and death. So, it becomes crucial to struggle to recognize the dignity of each person within the patient [5, 6], reflecting deeply on the fact that a human being is a biological, psychological, social and spiritual entity, and perhaps it is not possible to fully accompany any patient on their journey without understanding the essence of our nature. The modern doctor needs to professionally distance themselves from the patient has normalised a dehumanization of the health sciences [7]. It is time to remember that "a good doctor treats the disease; while the great physician treats the patient who has the disease"(William Osler) [8]. Medicine exists because of the patients who, with their individual experience, participate in reality, with their personal history and with a personal sense of transcendence, and reflecting on these aspects is critical to truly promote and experience medical vocation.

The COVID 19 pandemic has revealed a crucial and sometimes forgotten aspect of the Doctor-Patient relationship [8]. Without doubt this is one of the aspects that has been most affected by the current epidemiological situation, with patients fighting for their lives but distanced from their relatives and only close to nurses and physicians, but not always. This chapter seeks to reflect on these aspects, portraying a new vision of the essential within the Doctor-Patient relationship; the spiritual connection.

Doctor-Patient Relationship

The integral understanding of the human being as a biological-psychological-social and spiritual entity is at the core of the concept of healing. Medicine must not be understood as a rigid branch of knowledge since the accompaniment and healing process is much more complex and dynamic, implying in itself both a wide scientific knowledge and a huge human aspect; dialogue, compassion, suffering and trust. And, above all, we should never forget that the human touch is crucial for a true healing process.

The anthropological concept of great philosophers and authors, from Aristotle [9], Saint Thomas Aquinas [10], Kant [11], Elio Sgreccia [6], and even Saint John Paul II [12], coincide in identifying 3 elementary faculties: (1) Will, (2) Intelligence and (3) Affectivity. These faculties allow the human being to satisfy the teleological aspirations imprinted in the depths of human nature. Man thirsts for eternity; since the beginning of medical history we have seen how Aesculapius defeated death with his caduceus, thus seeking to satisfy the desire for eternity. The human being strives for transcendence, which is only achievable through the holistic experience of the previously described faculties. Man is capable of knowing truth as a scientific construct [12]. Man is capable of choosing and acting well based on natural law [12], and, above all, human beings are capable of achieving love [12] by acting in accordance. The spiritual essence is encountered in the unsatisfied teleological desire; medicine itself arises from the need to transcend time, matter and history [13]. When one human being sees another human being suffering in misery and decides

Fig. 6.1 Integrative anthropological dimensions, by Ana Lucía Soní Dillmann

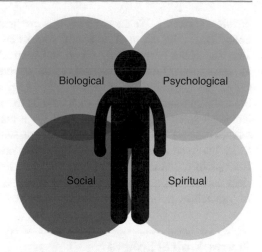

to help and accompany them, he/she is demonstrating the pure teleological manifestation of the Healthcare profession.

Throughout the Pandemic, we have seen how health professionals have based their work on these three faculties (Will, Intelligence and Affectivity); working arduously, and at times forgetting about the enormous personal sacrifice [14]. How can we understand this sacrifice if not as a form of love of humanity?

The spiritual nature of the medical vocation and of the doctor-patient relationship is fundamentally understood through its cornerstone: the dignity of the person [15]. The immeasurable value of the human being which inspires and aspires to protect life in all its senses, allows us to transcend, thus -medicine- is both a humane and a humanizing act.

Although not frequently discussed, I believe that the Doctor-Patient relationship is based on the spirituality of human nature. Dignity unites all humanity in spite of racial, religious, gender, and cultural differences, and what distinguishes the individual in each human being is our spiritual essence.

> I solemnly swear to dedicate my life to the service of humanity
>   (Geneva Convention) [1]

The famous Hippocratic Oath [1] shows the necessary and implicit dedication to the profession, dedicating life to the service of humanity in all its spheres, thus promoting a comprehensive vision of our patient (Fig. 6.1).

Biological Sphere

> Do you want to be a doctor, my son? Aspiration is this of a generous soul, of an avid spirit of science. Do you want men to consider you for a god who alleviates their ills and drives away fear? (Aesculapius) [4].

The biological integrity of the human being is one of the elementary values on which professional medical practice is founded; thus, doctors exist to alleviate physical suffering. We cannot deny this integral essence, since the human being, by definition, is the union of body and soul in an individual and indivisible way [10, 16]. The value bestowed by corporeality speaks of the spirituality imprinted into human nature. How many diseases of the psyche are manifested somatically?SARS CoV2 has highlighted once again our biological fragility; generating reflection on the predominant value of the 4 spheres in the integration of the human being. The technical-scientific domain is crucial for a correct professional practice since we cannot pretend to diagnose and treat anyone without first knowing in depth the mysteries involved in human biology. However, unfortunately in the XXI century there is a risk of reducing man to mere biological reality [17], and for this we must continuously reflect on why and how medical service must be centred around the love for our fellow human.

Psychological Sphere

> The human being is an animal of rational intelligence.
>   (Boecio) [18]

Man as an indivisible individual cannot be understood without referring to the psyche; its complexity refers to the thirst for eternity. It is not possible to isolate biology from the psychological reality of the human being since we are biology complemented and shaped by a psychology that must be understood as an integral and integrating part of our nature. This is where medical practice becomes more and more complex since it is not only physical or even technical. When the doctor finally meets the patient, the practice of medicine needs to take into consideration such crucial aspects as emotional, cognitive and interpolative integrity. Medicine and medical education face great challenges in this regard [19]; challenges that the epidemiological reality has highlighted. The development of "soft" competencies is increasingly necessary, so that archaic models such as paternalism can be transformed into a much more integrated and collaborative doctor-patient relationship [20], which highlights patient autonomy, dignity and individual value.

Social Sphere

> No man is an island.
>   (John Donne) [21]

Human beings are gregarious by nature, as humans need society, and the sense of belonging is perhaps one of the most interesting characteristics in human nature, and the social distancing that we have experienced recently has undoubtedly stressed this reality [22]. Medicine must be developed and practiced in this sense. It must be clear that the participation of doctors is crucial in the harmonious development of a society strengthened by anthropocentric senses [23], where the person is the true, definitive value and end of all social actions. Perhaps never before have we understood the prevailing need for collaborative, interdisciplinary and cooperative work in our globalized reality. This need is fundamental for us to re-emerge from a crisis, such as the current one, strengthened with those elementary values needed for the development of life in society.

Spiritual Sphere

> I solemnly promise to watch over human life with the utmost respect.
> (Geneva Convention) [1]

This is perhaps the least studied of scientific reality. However, it is critical for understanding how we experience human reality. We have already delved into some characteristics such as the teleological aspiration and dignity of the human person in medical practice, however, we can safely assert that the process of self-knowledge is in itself a life-long process.The doctor-patient relationship -as the basis of medical practice- is a humane and humanizing act, in that it unconsciously reveals the spiritual truth of man, which makes the bond engendered in the medical act palpable.

The current epidemiological situation which has paralyzed the entire world, will hopefully allow us to reflect on the vital importance of basing the doctor-patient relationship on fully humane criteria. Archaic models that do not promote the dignity of the person must in turn give way to increasingly humanized models. The loneliness and isolation that both our patients and our colleagues have experienced [22] cry out for the need to recentre our professional paradigms on the person as the guiding criterion.

The general consensus around the prevailing need to rehumanize the medical sciences [24] will be incomplete if we do not integrate all the component spheres of anthropological reality. We cannot deny the spiritual essence of human nature; on the contrary, we must be aware of the high degree to which the medical act contributes to a full experience of spirituality manifested in the thirst for eternity that is engendered in the dignity of the person. Human spirituality transcends religious and cultural characteristics, and we must be aware of this.

> I will not allow considerations of age, illness or disability, creed ... to come between my duties and my patient
> (Geneva Convention) [1]

Patient Spirituality

For us to fully understand this spiritual connection, we should also analyse and study the spiritual reality in our patients and in ourselves as eventual patients since it is undeniable that sickness is a path all humans will sooner or later have to cross. However, it is a much more complicated situation because in this sense we should not only comprehend the human nature related with one profession, but we should also understand it as the whole of human life.

Suffering (sickness) is an essential element of life, as Victor Frankl wrote in his book *Man's Search for Meaning* [25], and pain allows us to grow and mature by facing ourselves -in this way- with a much broader reality of "transcendence". Throughout history philosophers have concluded that tragedy (sickness, pain, suffering…) is the true source of personal humanization and solidarity [26], and by facing this reality we can glimpse our neighbour's experience, which will give us the opportunity to actively participate in the construction of a more harmonious society through the living exercise of empathy and generosity.

Unfortunately, twenty-first century hedonism, relativism and superficiality have distorted the comprehension of this crucial aspect, and the ideological preponderance for the rejection of pain has led us to the dehumanization of this sense. Suffering and Happiness are just two different sides of the same coin (life), there cannot be one without the other. At this moment, it is important to establish that, with these words, I do not wish to add "exalted" value to suffering, but only emphasise that we must know how to face and experience it as a complementary part of our lives. We must learn to perceive pain as an ordeal that may allow us to continue in the constant work of improving our human nature.

As stated we (humans) have been given three great tools, Will, Intelligence and Affectivity, and only through them we will be able to satisfy our teleological desire. This is ample reason why we should be compelled to live with them in wellness and sickness since it is only then that we will complete (in a more rounded way) our vision of life, and find meaning to one of the unsolved mysteries of human history: our own existence.

Practice Time

Since the experience of spirituality is of a personal and individual character, the following questions are proposed as a concrete means to reflect - at any time - on the foundation of the medical vocation "human contact" person to person, heart to heart.

The reader may find in the following section a space that favours inner knowledge and the deepening of the unexplored reality, but undoubtedly lived on a daily basis.

Physicians

- Why did I study medicine?
- Have my reasons changed with time?
- What is the current status of love for the profession?
- Do I live the Hippocratic Oath in my daily practice?
- Have I lost the person within the patient?

- Why and Where?
- Do I thirst/wish for transcendence?
- Do I transcend space and time with each of my patients?
- Do I fulfil my thirst for eternity?
- What am I missing?

Patients

Have you experienced true suffering in my life? (pain may be "physical, psychological or social and spiritual").
- How have you experienced this suffering?
- Were you alone?
- Have you found an opportunity in this suffering?
- Have you found meaning through pain?
- Have you had the opportunity to accompany someone in suffering?
- How was that experience?
- How do you feel about the reality of pain and happiness being two different sides of the same coin (life)?
- Can we transcend suffering?
- What are you missing to find full meaning in your life?

Conclusion

Love for medicine shows a great love for humanity, in all its aspects. The COVID 19 Pandemic and the reductionism inherent in the twenty-first century possibly presents the greatest challenge to medical education and practice. For this reason, I believe that we must focus on the prevailing need to re-humanize the healthcare sciences.

The spiritual connection in the medical act is undeniable; the aspiration and the thirst for transcendence are intrinsically manifested examples. If we are able to promote this characteristic we will in turn find an enormous sense of belonging and teleological satisfaction that will contribute to the harmonious process of social construction within the medical profession. The struggle between life and death will cease to be accidental and will take on a greater meaning in the defence of human dignity, which in turn will revive the love that is inscribed in the heart of every spirit that generously wishes to dedicate their life to the service of humanity.

> Your office will be for you a robe of Nessus. In the street at banquets, in the theatre, in your own home, strangers, your friends, your relatives will tell you about their ills ... your life will no longer belong to you ... Think about it while you have time. But if indifferent to ingratitude, if knowing that you will see yourself alone among human beasts, you have a soul stoic enough to satisfy you with the duty fulfilled without illusions, if you judge yourself paid enough with the happiness of a mother, with a face that smile because he no longer

suffers, with the peace of a dying man from whom you hide the arrival of death; if you yearn to know man and penetrate all that has been brought about by his destiny, then become a doctor, my son (Letter from Aesculapius to his Medical Son) [4].

Acknowledgments P. César Hernández Rendón, LC (Vice Rector, Universidad Anáhuac México).

P. Antonio María Cabrera Cabrera, LC. (Dean Bioethics Faculty, Universidad Anáhuac México).

Jonathan McFarland (Associate Professor, Universidad Autónoma de Madrid).

References

1. de Hipócrates J. Fórmula de Ginebra. Asociación Médica Mundial: Asamblea 8/11-IX–19.
2. Reddy PC. Post COVID-19, life will never be the same again [Internet]. The Hindu. 2020 [cited 2021 Mar 15]. Available from: https://www.thehindu.com/news/cities/chennai/post-covid-19-life-will-never-be-the-same-again/article31334665.ece.
3. Fins JJ, Edmund D. Pellegrino, MD 1920–2013. Transactions of the American Clinical and Climatological Association; 2015. p. 126.
4. Jones KB. The staff of asclepius: a new perspective on the symbol of medicine [Internet]. Wisconsin Med J. 2008. [Cited 2021 Mar 15]. Available from: https://www.researchgate.net/publication/5281453_The_staff_of_Asclepius_a_new_perspective_on_the_symbol_of_medicine.
5. Bermejo MP, Sgreccia E. The debate in bioethics. The personalism in the thought of Elio Sgreccia. Interview with Elio Sgreccia. Medicina y ética: Revista internacional de bioética, deontología y ética médica. 2018;29(1):159–74.
6. Sgreccia E. Persona Humana y Personalismo. Cuad Bioet. 2013;24(1):115–23.
7. Haque OS, Waytz A. Dehumanization in medicine: causes, solutions, and functions. Perspect Psychol Sci. 2012;7(2):176–86.
8. Young P, Finn BC, bru Etman JE, Buzzi AE. William Osler: El Hombre y sus descripciones [Internet]. Conicyt.cl. 2012 [cited 2021 Mar 15]. Available from: https://scielo.conicyt.cl/pdf/rmc/v140n9/art18.pdf.
9. Baracchi C. Aristotle on becoming human. Tóp rev filos. 2013;(43):93.
10. Aquino S. Summa Teológica. 2nd ed. Madrid: Biblioteca de Autores Cristianos; 2016.
11. Kant I. Antropología en Sentido Pragmático. Madrid: Alianza Editorial; 1991.
12. López AFL. Karol Wojtyla and his personalistic vision of man. [Internet]. Cuestiones Teológicas. 2012. [cited 2021 Mar 15]. Available from: http://www.scielo.org.co/pdf/cteo/v39n91/v39n91a07.pdf.
13. Scheler M. El puesto del hombre en el cosmos. 20th ed. Buenos Aires: Losada; 1994.
14. Hay que proteger el personal sanitario, héroes que luchan contra el coronavirus. 2020. Available at: https://news.un.org/es/story/2020/03/1472062. Accessed 18 Mar 2021.
15. Burgos Velasco JM. ¿Qué es la bioética personalista? Un análisis de su especificidad y de sus fundamentos teóricos. Cuad Bioet. 2013;24(1):17–30.
16. Garrocho Salcedo DS. El hilemorfismo en evolución. Una aproximación moral a la relación entre el cuerpo y el alma en Aristóteles. Universitas Philosophica. 2016;33(67):165–81.
17. Solana Ruiz JL. Reduccionismos antropológicos y antropología compleja. Gazeta de Antropología. 1999;15(08)
18. Boecio. La consolación de la Filosofía. Clásicos de Grecia y Roma. Buenos Aires: Alianza; 1998.
19. Seritan A, Hunt J, Shy A, Rea M, Worley L. The state of medical student wellness: a call for culture change. Acad Psychiatry. 2012;36(1):7–10.

20. Mendoza A. The doctor-patient relationship, bioethical considerations. Rev- Peru Ginecol Obstet. 2017;63(4):555–64.
21. Murn M, Graefe M. "No man is an island" -Crossing thresholds: journeying with the recent poetry of Syd Harrex. Trans Literat. 2016;9(1)
22. Zamarripa J, Delgado-Herrada M, Morquecho-Sánchez R, Baños R, De La Cruz-Ortega M, Duarte-Félix H. Adaptability to social distancing due to COVID-19 and its moderating effect on stress by gender. Salud Ment. 2020;43(6):273–8.
23. Domínguez Martínez J. Derecho civil: parte general, personas, cosas, negocio jurídico e invalidez. 8th ed. México: Editorial Porrúa; 2000.
24. Cleary M, Visentin D, West S, Lopez V, Kornhaber R. Promoting emotional intelligence and resilience in undergraduate nursing students: AN integrative review. Nurse Educ Today. 2018;68:112–20. https://doi.org/10.1016/j.nedt.2018.05.018.
25. Frankl V. El hombre en busca de sentido. Barcelona: Herder; 2016.
26. Callabed J. Curar a veces, aliviar a menudo, consolar siempre. 1st ed. Barcelona: Ediciones y publicaciones universitarias; 2011.

Covid 19 and Butterfly Thinking in the Flow of Time

Jonathan McFarland

Fig. 7.1 Photo by Jonathan McFarland—9th June 2020

J. McFarland (✉)
Faculty of Medicine, Autonomous University of Madrid, Madrid, Spain

Sechenov University, Moscow, Russia

© The Author(s), under exclusive license to Springer Nature
Switzerland AG 2022
M. G. Marini, J. McFarland (eds.), *Health Humanities for Quality of Care in
Times of COVID -19*, New Paradigms in Healthcare,
https://doi.org/10.1007/978-3-030-93359-3_7

Introduction

Time present and time past
  Are both perhaps present in time future,
  And time future contained in time past.
  If all time is eternally present
  All time is unredeemable. [1]

In 1972, at the 139th meeting of the American Association for the Advancement of Science in Washington D.C, Edward Lorenz asked the following question, "Does the flap of a butterfly's wings in Brazil set off a tornado in Texas?" [2]. Nearly half a century later this butterfly effect has changed the world, and the ensuing story is known to all. On January 9th, 2020, the Chinese CDC (Centres for Disease Control) reported that a novel coronavirus was the causative agent of a local outbreak of a cluster of pneumonia cases of unknown aetiology [3]. And then it all started, with one small ripple originating in Wuhan somehow transforming itself into a tsunami; the butterfly effect at its most horrifyingly effective. "For better or worse, the entire world is globally interconnected", [4] or, as Fichte wrote in 1800, *"You could not remove a single grain of sand from its place without thereby … changing something throughout all parts of the immeasurable whole."* [5] *Twenty-first-century life has had its pause button pushed but will the reset button follow suit?*Lepidopterology[a] has always had strong links with the medical sciences, and mine are also personal as I remember going to see Cyril Clarke, the distinguished physician and lepidopterist, lecture on Butterflies at the Liverpool Medical Institution[b]. Weatherall referring to Clarke's interest, states "Late in his career, and stimulated by his work on the genetics of mimicry in butterflies, he became interested in the evolving field of medical genetics" [6]. He is remembered as a leading researcher of the Rhesus haemolytic disease, which, until his discoveries, had been the cause of many stillbirths; my mother being one of the many women affected. So, my family history is tenuously threaded into this story. But butterflies have not only interested scientists or physicians but others, including two British prime ministers, Neville Chamberlain and Winston Churchill who were both amateur enthusiasts. Indeed, Churchill began to take an interest in butterflies at the age of six, writing to his mother "I am never at a loss to do anything while I am in the country for I shall be occupied with 'butterflying' all day (as I was last year)" [7]. He collected butterflies when in the army in the Sudan and India, and later constructed a butterfly garden at his home in Chartwell, Kent. Ten years ago the National Trust[c] renovated his butterfly garden. Nowadays, however, as mentioned in a recent newspaper article, [8] the commitment has shifted from collection to conservation with one of the most seminal books of the 1960s, *The Collector* by John Fowles[d], [9] mentioned as being partly responsible for this.

However, my interest is in another lepidopterological metaphor, rooted in another writer, Vladimir Nabokov, the renowned Russian novelist, also a distinguished butterfly collector and amateur scientist who received posthumous recognition for his migratory theories of the *Polyomatus blue.*[e] "At the end of a 1945 paper on the group, he mused on how they had evolved. He speculated that they originated in Asia, moved over the Bering Strait and headed south all the way to Chile" [10] (Fig. 7.1).

But what is Butterfly thinking? And how does it relate to this chapter? A recent book in the School of Life series explains how, "The truly precious thoughts have something almost airborne about them, so inclined are they to flit away at the slightest approach of our conscious selves" [11]. It mentions how many of the world's great thinkers have always likened thought and ideas to winged creatures, citing two examples, Plato who compared the mind to birds circulating within a large cage and Nabokov, where ideas are seen as butterflies, "we must learn to patiently wait until they can be coaxed into flying into the net…".

I will use this analogy to try to *net* some of the implications of the COVID pandemic on our lives, with words for butterflies guiding us like Virgil[f] leading Dante through *his* own hell in The Divine Comedy. It is possible that the pandemic has distorted our notion of time, heightening and intensifying it, and thus, in the words of Eliot, making "all time eternally present" [1]. For this reason, the chapter will be divided into three sections; **Time Present (In-Covid age)**, **Time Past (Ante-Covid age)** and **Time Future (Post-Covid age)**.

"And art thou then that Virgil, that well-spring,   *75*
From which such copious floods of eloquence
Have issued?" I with front abash'd replied.
"Glory and light of all the tuneful train!
May it avail me, that I long with zeal
Have sought thy volume, and with love immense   *80*
Have conn'd it o'er. My master thou, and guide!
Thou he from whom alone I have derived
That style, which for its beauty into fame
Exalts me" [12].

Time Present (In-Covid Age): Freedom and Normality

So many roads, so much at stake
   Too many dead ends, I'm at the edge of the lake
   Sometimes I wonder what it's gonna take
   To find dignity [13].

Now it is the beginning of May 2021. I sit in my garden, listening to the sound of our resident blackbird singing, and even attempting, but failing, to communicate with it using a new app on my mobile. Steven Lovatt writes, "While each bird species possesses its own distinctive calls and songs. In the blackbird the variety of sounds is astonishing", [14] and McCartney sung

Blackbird singing in the dead of night
   Take these broken wings and learn to fly
   All your life you were only waiting for this moment to arise [15].

Maybe the bird is trying to tell me something? Maybe the blackbird (like the phoenix) will help us arise from the ashes? Sitting and listening to birdsong I am one of the lucky ones; my family and I unscathed to date by the virus, but I am however strongly aware of the immeasurable distress and sorrow it has caused across the world. And this distress has been increased by a pronounced lack of unity in the handling of the pandemic; highlighted by the fact that whilst in some occidental countries the easing down on restrictions has begun, in other countries COVID 19 is still wreaking destruction. Disharmony and disunity prevail.

Evocative of Dante's decline into the underworld, India is being badly hit, with hospital admissions relentlessly increasing to such an extent that on the 26th of April India set a global record of 3352,991 cases a day, and on Saturday 8th May newspaper headlines read - *India records almost 4200 Covid deaths in a day* [16].

Zarir Udwadia, a consultant physician from Mumbai, writing in the Financial Times likens the COVID wards of India to scenes from Dante's inferno; "row upon row of patients waging desperate struggle to breathe, their cries for help often falling on deaf ears as overworked medical staff struggle just to keep going. Essential drugs are not in stock and most frightening of all, oxygen, that very essence of life, is in short supply" [16]. Arundhati Roy, Indian novelist, writing in the *Guardian*, states, "Oxygen is the new currency on India's more bad news Stock Exchange. Senior politicians, journalists, lawyers – India's elite- are on Twitter pleading for hospital beds and oxygen cylinders" [17]. This is the current situation in the world's largest democracy. The virus does not distinguish between rich and poor, young and old; it is after all only a virus; and moreover, there is sustained debate as to whether viruses are living organisms or not. However, it has shown again and again that it is more concentrated, more severe and more deadly amongst the least favoured who live stacked together in squalid conditions with little access to diagnostic testing or treatment. And vaccines, instead of bringing the world together, are testimony to an even greater rift between countries and peoples and cultures, as captured in a recent article in Spanish titled *El apartheid de 2021: las vacunas* (The apartheid of 2021: vaccinations) [18].

Where is the equality? The unison in response? When villages hardly able to cope with easily treatable diseases such as diarrhoea and tuberculosis are inundated by returning migrant workers escaping the big cities and bringing with them the disaster of COVID-19. Roy continues, "In the UK, vaccines are free and a fundamental right. Those trying to get vaccinated out of turn can be prosecuted. In India, the main underlying emphasis of the vaccination campaign seems to be corporate profit" [17]. And this happening in the country which is one of the world's main vaccine producers.

So where does this leave humanity? Where in the words of Dylan has our dignity gone?

And can we ever unmask it again? And what about words, when they begin to be used so lightly that they lose their meanings or they are, purposedly or not, misused? It was Rudyard Kipling[g] who, on addressing surgeons in 1923, warned that "words the most powerful weapon known to mankind" [19]. But what happens when words lose their validity because they have been distorted or perverted through an apparent lack of either precision or truth, seemingly ubiquitous in twenty-first century life or at least those claiming to represent and make decisions for us? Words have consequences and cannot be bandied around like confetti at a wedding, colourful and at best ornamental but at worst, corrosive and destructive. Think carefully before you use words like **freedom** and **normality**, and more so, if you are being listened to by an entire country, with your words influencing the actions of many.

Freedom is a word that is being wielded around with great force by many politicians around the world; think Trump, Bolsanaro, Lopez Obrador. They use the word however unremittingly without first paying attention to what it really means and secondly without anticipating and heeding the possible consequences. The latest in this distinguished line of prominent politicians to use the word is Diaz Ayuso, a politician aligned to the far-right who was recently elected president of the community of Madrid and is responsible for governing around four million people. She won re-election centred on one word "freedom", and on winning the elections, she hailed the result as "another triumph for freedom in Madrid" [20]. But what does she mean by freedom? How can people use this word, when others are dying in the ICU in hospitals around the region? or when health care professionals have risked both their physical and their mental health? And continue to do so? I do not believe that she was talking about the freedom that Nelson Mandela spoke about when he said, "For to be free is not merely to cast off one's chains but live in a way that respects and enhances the freedom of others" [21]. This word was rooted in the struggle for racial equality, amongst others, which does not give you the liberty to do all you want, as Diaz Ayuso has claimed. This perversion of language was seen a few days later, when thousands took to the streets in Madrid and other cities around Spain, celebrating the end of the nationwide curfew, unmasked, carousing in large groups and singing the praises of freedom. But words as Kipling knew, have consequences; and those that utter them, responsibilities.

In this new clamour for a return to normality perhaps we are missing the point. Perhaps we should not be looking to go back to a return to normality but looking to go forward, to push forward to new paths and new ways to the waterfall? [22]. Slavoj Zizek writes, "the COVID-19 pandemic shattered not only our health care, economy, political and social relations, and mental health, it did something much more radical to which only philosophy can give access: it threatened our sense of "normality," a term we must interpret in all its weight" [23]. He then goes on to explain how normality stands for what Lacan called the "big Other", which structures not only our inner life but also how we relate to what we see and live as "reality". A reality, our normality, is not the same now and will never be the same again; just as we listened to the leaders and generals at the beginning of the First World War with their propaganda of 'it'll be over by Christmas' we are listening to this shriek or cry for the new normality. Propaganda indeed, for as Zizek notes, "Instead of dreaming about to return to the old normality we should engage in a difficult and painful process of constructing a new normality" [23]. Richard Horton writes, "what matters is that he Zizek is asking questions should be central to our public discussion about a pandemic and our post-pandemic future" [24].

The inequality of the virus leads me to ponder the term syndemic, which Horton mentions in his article in The Lancet. "Syndemics are characterised by biological and social interactions between conditions and states, interactions that increase a person's susceptibility to harm or worsen their health outcomes" [25]. The term that highlights the exacerbations of the inequalities of any disease is derived from the ancient Greek language and introduced in the 1990s by Merrill Singer, an American Medical anthropologist. Horton thus is looking at COVID-19 as not "just another infectious disease" since "COVID 19 is not Black Death" but as a "disease killing mostly disadvantaged and fragile people, Without identifying and without intervening on the cofactors making SARS-Cov-2 a lethal virus, no measure will be truly effective, not even vaccines" [25].

The present is also called the *now*, but what exactly is now? What is the present? In the mid-twentieth century, the Pop artist, Andy Warhol along with photographer, Nat Finkelstein, coined the term "15 minutes of fame". Andy Warhol purportedly stated that everybody wished to be famous, and Finkelstein retorted, "yeah, for 15 minutes" [26]. This riposte is a pertinent portrayal of how rapidly time was seen to pass in the late 1960's. A reflection of what Zygmunt Bauman termed *liquid modernity*,

> What all these features of fluids amount to, in simple language, is that liquids, unlike solids, cannot easily hold their shape. Fluids, so to speak, *neither fix space nor bind time*…….fluids do not keep to any shape for long and are constantly ready (and prone) to change it; and so for them it is the *flow of time that counts*, more than the space they happen to occupy: that space, after all, they fill but '*for a moment*' [27].

And now, I believe, that, as we enter the third decade of the twenty-first century, the "fluidity" or "liquidity" of time has intensified, with its parallel reduction in human attention span, to such an extent that the 15 min of the 1960s could be cut to seconds or, in keeping to a viral scale, to nanoseconds. But what about the Ancients?

Time Past (Ante-Covid Age): Mimesis and Solitude

> Words move, music moves
>   Only in time, but that which is only living
> Can only die. Words, after speech, reach
> Into the silence [1].

In 1905, in a small city in Switzerland, one of the most important discoveries of that century or perhaps any century was made. A young clerk went to meet his best friend to wrestle with a riddle that had been obsessing him for more than a decade; they spent two days analysing the ideas and trying to get to the core of the dilemma. They supposedly failed because the young man shuffled off home to his family to rest and sleep. The next day however he returned rejuvenated and triumphant; a changed man, and without even greeting his friend, he blurted out, "Thank you. I've completely solved the problem," went home and spent six weeks writing up the theory of relativity. Olivia Fox Cabane and Judah Pollack use this example of Albert Einstein as a way of introducing breakthrough (or Eureka) ideas, remarking that, "Every single person on the planet has the ability to generate creative breakthroughs inside their brains……..we all have a natural ability to create *butterflies*. [But] …You need skills and practice to build and wield a net" [28].But why are butterflies equated with this kind of thinking, and why, more specifically have I used this metaphor in this chapter? We need to go back to find the answer. According to an online etymological dictionary, "butterfly is the common name of any lepidopterous insect active in daylight, from the old English butterfleoge, evidently butter (n) and fly (v), but the name is of obscure signification." It continues that arguably this has to do with its colour or even an old notion that they actually ate milk or butter. But nobody really knows. The butterfly is often associated with a frivolous person, or with the nerves before an exam or even the feeling of falling in love as well as a swimming stroke first introduced into the English language in 1935 [29].

But it is the Greek word for the butterfly that takes us deeper, "The word for butterfly in formal Greek is *psyche*, thought to be the soul of the dead". Ancient Greeks also named the butterfly *scolex* ("worm"), while the chrysalis – which is the next stage of metamorphosis from a caterpillar – was called *nekydallon*, meaning "the shell of the dead" [30]. It is intriguing to learn that the butterfly for the Greeks was the representation of the soul or the spirit, but we can catch this connection throughout history; this union between beauty and science that is commonly represented by the butterfly. Even the so-called father of modern neuroscience, Santiago Ramon y Cajal, Nobel prize winner in Physiology for his investigations into human neuroanatomy, as well as a fine artist, wrote in his autobiography, "…Like the entomologist hunting brightly colored butterflies, my attention was drawn to the flower garden of the gray matter, which contained cells with delicate and elegant forms, the mysterious *butterflies of the soul*, the beating of whose wings may someday clarify the secret of mental life …." [31]. The similarities to Nabokov's fascination with butterflies, "I discovered in nature the nonutilitarian delights that I sought in art. Both were a form a magic…." is compelling, but Nabokov continues, "… both were a game of intricate enchantment and deception" [32], which is essentially an explanation of *WHY* butterflies and "butterfly thinking" are interwoven, intertwined, and perhaps more sinisterly entangled with COVID-19, indeed with any virus.

Both butterflies and viruses are experts in the art of deception, magicians; "Art is a deception that creates real emotions - a lie that creates a truth. And when you give yourself over to that deception, it becomes magic" [33]. Butterflies use a strategy called crypsis, where they reveal the undersides of their wings to blend in with their surroundings to protect themselves, while viruses, with coronavirus a particular expert, hack into a host's body's immune system in order to propagate successfully within that body, be it animal or human [34].

Thus, art, deception and magic usher in my next word, **mimesis**. Mimesis comes from the Greek word mimēsis meaning imitation, though in the sense of "re-presentation and not "copying"; with both Plato and Aristotle referring to it as a re-presentation of nature". For Plato, artists are imitators of an imitation, twice removed from the truth, which was God's creation. Shakespeare uses Plato's and, perhaps more precisely Aristotle's interpretation of tragedy, where man falls from a higher to a lower position. So, art imitates or mimics nature or the action of life, as to some extent, both butterflies who mimic to save themselves from predators and a virus that mimics to live and thrive, do. "The purpose of playing, whose end, both at the first and now, was and is, *to hold as 'twere the mirror up to nature*: to show virtue her feature, scorn her own image, and the very age and body of the time his form and pressure" [35]. Koonin and Starokadomskyy write,

> The history of life is a story of parasite-host coevolution……. All organisms are communities of interacting, coevolving replicators of different classes. A complete theory of replicator coevolution remains to be developed, but it appears likely that not only the differentiation between selfish and cooperative replicators but the emergence of the entire range of replication strategies…….is intrinsic to biological evolution [36].

Humans or homo sapiens are social creatures as Aristotle states. *"Man is by nature a political animal"* [37]. However, the last year has taught us that perhaps we also need to learn how to be alone and live in **Solitude**. "Know thyself" is one of the ancient Greek aphorisms, more precisely one of the Delphic maxims; the first of three written in Apollo's temple at Delphi according to Pausanias [38], which, along with the other two, "nothing to excess" and "surety brings ruin" could well be seen as the maxims of a good doctor or medical practitioner. The latter maxim interestingly echoes the times we are living through with this deep-seated lack of "surety"; this in-grained and abrupt uncertainty. However, even though most see this as destructive, in **Time Future** I will argue to search for the positivity within uncertainty.One of the most interesting periods of Christianity is that of the desert fathers, early Christians who relocated to the deserts of Egypt from around the third and fourth Centuries AD, and one of the most interesting of the desert fathers was St Anthony the Abbot (also called Anthony the Great or the "Father of all Monks") who "achieved the serenity in adversity which is the lesson of desert monasticism. Peace of mind comes through solitude and fortitude" [39]. I maintain that, although present-day solitude has been imposed on us by COVID 19 and the early Christian fathers forced it on themselves voluntarily, there are still many lessons to be learnt from these early Christians. Even though we are primarily social animals, being alone or 'in communion with ourselves' to uncover an interior silence (and perhaps peace) is at times, not just an interesting option but indeed the only option. Finding a moment of solitude to reflect on *why* we are here and *what* we are doing is interesting, even critical, and may conceivably help us understand the words of Sogyal Rinpoche,[h]

> Reflect on this: The realisation of impermanence is paradoxically the only thing we can hold onto, perhaps our only lasting possession. It is like the sky or the earth. No matter how much everything around us may change or collapse, they endure [40].

Human beings are insignificant, and this is fundamental to remember from time to time.

Time Future (Post-Covid Age): Uncertainty and Ancestor

> The Sex Pistols, strictly speaking, were right: there is no future, for you or for me. The future, by definition, does not exist. 'The Future' whether you capitalize it or not, is always just an idea, a proposal, a scenario, a sketch for a mad contraption that may or not work.

'The Future' is a story we tell, a narrative of hope, dread or wonder. And it's a story that, for a while now, we've been pretty much living without [41].

The future is a story, but the question is how will we be telling the story of the COVID-19 decade, in 2030 or in 30 years? And, still left in my butterfly net, I have two words—**Uncertainty**, and **Ancestor**. At school, I directed Tom Stoppard's brilliant play *Rosencrantz and Guildenstern are Dead* [42], an invention of what happens to these secondary characters when they are not on stage in Shakespeare's *Hamlet*. It is a sharp, tragicomedy often compared to Beckett and especially his *Waiting for Godot* since the two main characters in both plays could be seen as two halves of one person. I was particularly fascinated by the opening where Rosencrantz and Guildenstern bet on coin flips, with the incredibly improbable outcome that Rosencrantz betting on heads wins 92 flips in a row. Carlo Rovelli in *There Are Places in the World Where Rules Are Less Important Than Kindness* explains, "in this essentially uncertain world, it would be foolish to ask for absolute certainty" [43]. Indeed. Ninety-two flips in the row does not exist; Certainty, or absolute certainty, does not exist, and he continues,"But that doesn't mean that we are completely in the dark. Between certainty and complete uncertainty there is a *precious intermediate space* – and it is this intermediate space that our lives and our thoughts unfold" [43]. The pandemic has made us more aware of this, since isn't it the balance or this *precious intermediate space* between the certain and the uncertain that is always crucial? Balance and equilibrium are everything, as always, I suppose.

Bruno de Finetti was an Italian physicist born in Innsbruck at the beginning of the twentieth century, whose life work was dedicated to the mathematical theory of probability. Rovelli in his article *Bruno de Finetti: Uncertainty is not the enemy*, writes

We cannot get rid of uncertainty. We can diminish it, but we cannot make it disappear. Hence, we should not experience it as some kind of nightmare. On the contrary, we should be reconciled to it as our lifelong partner [43].

And he goes on to explain how it [uncertainty] is what brings interest to our lives, what makes us experience new things, both good and bad, and what makes life truly worth living. So, instead of scorning and shunning the uncertain, we should welcome it with open arms. The pandemic has strengthened this, reinforced the importance of the uncertain; most have seen it as a negative, but I follow Rovelli and the title of his above-mentioned essay, uncertainty is *not* the enemy. The prior belief that certainty was the key to all; now that way of thinking was the true enemy. This reminds me of Siddhartha Mukherjee in his short book, *Laws of Medicine: field notes from an uncertain science* [44], since both de Finetti and Mukherjee, with nearly a century between them, highlight the importance of Thomas Bayes'[1] work where probability is explicitly rooted in individual experience. "There is no absolute knowledge. History repeats itself- and so do statistical patterns. The past is the best guide to the future" [44]. Medicine and the world around us need to embrace uncertainty which is central to all.As mentioned earlier vaccines instead of bringing humanity together are seemingly dividing it. This was not always the case. Jonas

Salk, the American virologist and medical researcher, when he developed the polio vaccine in 1955 was hailed as a "miracle worker" [45]. However, when asked how he would wish to be remembered, he answered that he would like to be known as the man who asked the question "Are *we being good ancestors*?". What is an **ancestor**? An ancestor is normally seen as a person further removed than a grandparent from whom we are descended.

However, interestingly enough it also refers to plants, animals and even machines. So, everything within the world we live in, the human, the vegetable, the animal and the man-made, is interconnected by this one word. Perhaps a reference to how entwined we are? Certainly, this could be seen as yet another example of the intrinsic globalization at the centre of twenty-first century life, and, for me at least, a sign that we should not be thinking in the short term (the minutes or seconds that make up our modern attention span or the trend for the instantaneous and instant) but to the long term. In this way, we will attempt to be the good ancestors that Salk yearned for. And, perhaps instead of always looking to the future; 'future trends', 'future fashion', 'future this, future that', we could take *'the now'* as being slightly longer than the proverbial blink of an eye. Now would refer to a *long now* that could be 100 years, or even a thousand, since maybe with this time scope we would *really* see what we are facing. A long-term approach, the long now, would encompass the human, the natural and the man-made worlds. We would be able to see that just as this pandemic was enacting a butterfly effect from the natural to the human, it is not by any means an isolated event. We are entangled with nature and need to look after the planet to look after ourselves; the sooner we understand this the better. Human health and well-being do not exist without planetary health and well-being.

A virus, a coronavirus if you like, is probably not even a living organism but is adept (like the butterfly) in mimicking other organisms. It does this to survive within animal and human bodies, and when the conditions are correct, to wreak havoc, damage and kill. It does not however kill for pleasure or to defend itself as we [humans] do on occasion, but for survival. But are humans so far removed from viruses? And who are we really? We are of course built of flesh, muscles, organs, blood etc.; our personality is confectioned by our DNA, our genetic tree of life but, since we are around 95% microbiota, how much of one human is made up of their own DNA? Put another way, we are not really ourselves but around 95% comprised of other organisms; other organisms from outside our own body. Invaders, like viruses. So, paradoxically, humankind by infringing into the natural world (rainforests, tundra, the virgin areas of the world) and impacting the fragile exterior ecosystems adversely, are also harming our own microbiota, our own interior ecosystem. By jeopardising the natural world, we threaten our own. We are self-destructing from without and from within. Humanity to survive needs to protect the natural world for we are doomed, if we do not, and we will have utterly failed Salk's test.

"A few millennia ago, the God of Genesis (whose sayings were set down before He fell silent) had been precise: Be fruitful, and multiply, and replenish the earth, and subdue it" [1, 28]. One might reasonably assume (no disrespect to the priesthood) that this project had been accomplished. The earth subdued, and that it was time to give the womb some respite. We were eight billion humans. There were only a few thousand snow leopards. Humanity was no longer playing fair" [46].

Perhaps it is the moment to start playing fair again, if indeed we ever have.

Conclusion: Stillness

> At the still point of the turning world. Neither flesh nor fleshless;
>   Neither from nor towards; at the still point, there the dance is,
>   But neither arrest nor movement. And do not call it fixity,
>   Where past and future are gathered [1]

This chapter begins and ends with references from T.S. Eliot's *The Four Quartets* [1]; both are relevant if not crucial to what we have been living through and what we are still living through. The first stresses the interconnectedness of time, the normal way of construing its movement, with the past influencing the present, which, in turn, is a steppingstone to a brighter and clearer future. It is linear and chronological and, dare I say, progressive. However, this pandemic has left us uneasily situated in, and connected to, time. Eliot's word "perhaps" gives a clue to this uncertainty wound up like a wristwatches' spring, as well as coupling conveniently with the "maybe" of the concluding chapter. Nothing is sure, much less time, for as Berger says, "*The flow of time is turbulent. The turbulence makes lifetimes shorter-both in fact and subjectively. Duration is brief – nothing lasts*" [47]. Even though some butterfly's lives last considerably longer, the duration of a butterfly's lifetime is normally around one month, but this depends on many factors, species, size, season and even location. Their time-flow is turbulent in that they metamorphize through four stages: egg, larva, pupa to adult; reminiscent to some extent of Shakespeare's *Seven stages of man* speech,

> All the world's a stage,
>   And all the men and women merely players,
>   They have their exits and entrances,
>   And one man in his time plays many parts.
>   His acts being seven ages [48].

During this last year and a half human beings have truly been *merely players*, with the main producer, director and star played by one living (or non-living) virus measuring approximately 80-120 nm in diameter. That minuscule star has taken *all* the credit for the turbulence caused, tearing up the normal rules and making us doubt everything. And time, as always, plays its part.However, perhaps this temporal uncertainty can be seen as a positive, the proverbial light at the end of the tunnel or silver lining to the cloud? This guides me to the last of the butterflies (words) **stillness** since time has that capacity, to move fast but remain still. In the second quote from Burnt Norton, Eliot refers to "*the still point of the turning world*". Perhaps we have reached that point? U2 caught the sentiment well with the lines, "*Running to stand still*" [49] as a speed-crazy world *did* appear to stop running in spring 2020, albeit momentarily when skies were empty of plane streaks, birdsong and not alarms woke us, goats ambled through Llandudno, Wales, and dolphins returned to a Venice free of tourism. Perhaps this was a necessity? A year ago, in April 2020, I wrote the following words

> I have always believed that silence is a pause, and one of the most treasured things in the frenetic 21st Century life; certainly, silence is the key to music and other arts. An opportunity to reflect on my situation, and that of my family [50].

Precious needed moments, resonating with others, "The need for an empty space, a pause, something we have all felt in our bones; it's the rest in a piece of music that gives it resonance and shape" [50], and more recently in *The Interior Silence* which ends thus, "Silent is an anagram of listen. It is how I shall try to live my life, as the monks taught me. Attentive to the interior silence" [39]. The rest was much-needed; paradoxically life-affirming in troubling and trying times, and maybe it was the moment to listen to the natural world around us, its cries for help, for shouldn't we first be looking to planetary health, and our own health, second? Coope, in a recent article, emphasizes this, explaining how senior figures from the UN and WHO "have drawn attention to relationships between human impacts on the natural world and human disease pandemics, suggesting that 'Coronavirus is a warning to us to mend our broken relationship with nature'" [51].One year has passed. The *pause* has come and gone. The man in the street, fuelled by the media, and *sometimes* aided by governments, begs to return to a normality that does not exist. But there are however obvious reasons, economic, social, psychological and mental for insisting that it may. Is it that really deep-down, "*humankind cannot bear much reality*"? [1].

Nevertheless, this prolonged and enforced pause did exist; made us think, sit up, take note, reflect and question the unrelenting flow of modern times and made us even think about or consider a new way of living. I felt and heard the silence and stillness one year ago, and even though it is quickly being drowned out again by incessant noise, it is still there, *just*. Resonating within. My tentative dream is that it may help us towards some kind of positive transformation, to grabbing hold of the hope that periodically appears "*like a flame in the darkness*" [52].

Isn't now the moment to hold that candle together and make a concerted effort to build towards a more **fair**, more **equal**, more **sustainable** and more **inclusive** future (Fig. 7.2)?

For are not the words of Benedetti, "*Y en la calle codo a codo somos mucho más que dos*" [53]—and out in the street arm in arm we are much more than two- true?

Postscript

On the 12th of May 2021, the following was published in an international and highly respected newspaper, "A swift international response could have stopped the 2019 Covid-19 outbreak in China becoming a global catastrophe in 2020, according to a scathing report on the response of world leaders and the World Health Organization to the pandemic" [54].

Fig. 7.2 Photo by
Jonathan McFarland on
29th April 2021

And it laid out the steps needed to prevent another, and probably worse, pandemic, stating among other things that the world's governments needed to succeed in attaining proper funding to be able to successfully stop the next pandemic in its tracks. The consequences of not being fully prepared for the next 'big one' are catastrophic to say the least, and whether the world's big economies will be ready to hand such powers to a body they cannot control [the WHO] will be a test of the extent to which the pandemic has rekindled multilateralism. But the urgency is clear. Another animal virus could jump the species barrier tomorrow. As the experts concluded last week, we have been warned.

Additional glossary (facts and figures indispensable to the telling of this story)

a	Lepidopterology (from Ancient Greek λεπίδος (scale) and πτερόν (wing); and -λογία -logia) is a branch of entomology concerning the scientific study of moths and the three superfamilies of butterflies
b	The Liverpool Medical Institution is a historic medical organisation based in Liverpool, United Kingdom. Its building on the corner of Mount Pleasant and Hope Street was opened in 1837, but the site has been used as a medical library since 1779
c	The National Trust for Places of Historic Interest or Natural Beauty, commonly known as the National Trust, is a charity and membership organisation for heritage conservation in England, Wales and Northern Ireland. In Scotland, there is a separate and independent National Trust for Scotland

d	*The Collector* is a 1963 thriller novel by English author John Fowles, in his literary debut. Its plot follows a lonely, psychotic young man who kidnaps a female art student in London and holds her captive in the cellar of his rural farmhouse
e	The common blue butterfly is a butterfly in the family Lycaenidae and subfamily Polyommatinae. The butterfly is found throughout the Palearctic. Butterflies in the Polyommatinae are collectively called blues, from the coloring of the wings
f	Publius Vergilius Maro, usually called Virgil or Vergil in English, was an ancient Roman poet of the Augustan period who wrote three of the most famous poems in Latin literature: the Eclogues, the Georgics, and the epic Aeneid. He was also Dante's guide through Inferno (Hell) in *The Divine Comedy*
g	Joseph Rudyard Kipling was an English journalist, short-story writer, poet, and novelist. He was born in India, which inspired much of his work. Kipling's works of fiction include The Jungle Book, Kim, and many short stories, including "The Man Who Would Be King"
h	Sogyal Rinpoche (1947–2019) was a Tibetan Dzogchen lama of the Nyingma tradition. He was recognized as the incarnation of a great Tibetan master and visionary saint of the nineteenth century, Tertön Sogyal Lerab Lingpa
i	Thomas Bayes was an English statistician, philosopher and Presbyterian minister who is known for formulating a specific case of the theorem that bears his name: Bayes' theorem
j	Chap. 11: *Maybe in, Maybe out, May be with the pandemic*
k	"And, in the Street, hand in hand// We are far more than two" (Translation by author)

> **Practice Time**
>
> Few would disagree that in the last a year and a half the world has been turned upside down, or at least our vision of it, with health care and all that it entails being at the centre of this uprooting. The Practice Time below appropriately focusses on how we could re-balance ourselves, on a possible homecoming. Read it to ponder on the ideas expressed and then reflect on the questions posed at the end.

COVID-19 Pandemic, Possible Condition for Homecoming

Negin Nouraei, Saba Mirikermanshahi, and Mahdi Azadi Badrbani

> Questioning Builds A Way
>   *Martin Heidegger*

Amid a Pandemic Crisis

Talking about the current pandemic crisis is inevitably accompanied by uncertainty and confusion. The different processes that, until recently brought order to our ordinary lives have all come to a halt or undergone fundamental changes. In other words, due to the circumstances that have arisen, what used to be considered normal is now in the spotlight. Concepts such as work, education, communication, and business have lost their common meaning; hence, each of these, while continuing differently in cyberspace, is not making sense as before. Since we cannot imagine an early end to this crisis, the lack of current knowledge to deal with the new disease and the daily exposure to evidence that shows the ineffectiveness of measures taken to control the current situation adds to the turmoil and uncertainty caused by this crisis [55].

The bulk of news related to COVID-19 shows the importance of the concept of health in this period. Statistics on the number of patients and casualties are updating rapidly and the latest findings in the field of public health are informing people. In addition, human health issues in this pandemic are becoming more prominent and the inability of the health system to understand and respond to these issues is evident. Also, the concept of health and disease has changed, and disease is no longer the only known condition for an individual; rather, anyone around us or even each of us may be a silent carrier of the disease. Therefore, our understanding of illness and health has disrupted due to the fear of getting sick and the possibility of imminent death becoming more tangible. In this article, by presenting Martin Heidegger's concept of "not-being-at-home" and based on Kevin Aho's interpretation, we try to describe the current crisis. Then by expressing the concept of illness and its similarity with "not-being-at-home," we try to revise the concept of health from Fredrik Svenaeus's standpoint. We believe that amid this pandemic is the proper time to reconsider the concept of health. This article tries to deepen our understanding of health and illness by discovering this concept through *Dasein*-based thinking, where *Da-sein* means being (*Sein*) there (*Da*). After highlighting this issue, we will point to the need to adopt a new approach in the modern health system to understand existential health and illness, considering the inadequacy of modern medicine in the face of human suffering and pointing to the more tangibility of this inadequacy in the current pandemic. Our approach in this paper is philosophical, and we explore health and illness on an ontological level.

Human as Dasein

Martin Heidegger[1] challenges the concept of a human being composed of mind and body by proposing the concept of Dasein. In Heidegger's fundamental text "*Being*

---

[1] Martin Heidegger (26 September 1889–1826 May 1976) was a German philosopher who is regarded as one of the most important philosophers of the twentieth century. He is best known for contributions to phenomenology, hermeneutics, and existentialism.

Published in 1927, Being and Time (German: Sein und Zeit) is standardly hailed as one of the most significant texts in the canon of (what has come to be called) contemporary European (or Continental) Philosophy.

This book put Heidegger to a position of international intellectual visibility and provided the philosophical impetus for a number of later programmes and ideas in the contemporary European tradition, including Sartre's existentialism, Gadamer's philosophical hermeneutics, and Derrida's notion of 'deconstruction'.

and Time", *Dasein* is introduced as a term for the specific type of being that humans possess. Heidegger uses *Dasein* for the being of humans to distinguish it from "subject". *Dasein* cannot be reduced to reason or soul, and it includes all human existence. *Dasein* is "being-in-the-world" in everyday life, which is always present in the world as a part of it. Heidegger considers this "being-in-the-world" to emerge from dwelling in the world and becoming familiar with the phenomena within it; everydayness is also an aspect of his/her "being-in-the-world", which means being immersed in the practical flow through life and acting in a conventional and pre-reflective way. "Being-with-others"_ *Mitsein*_ and living in a sharing world with them and concerning them is one of the fundamental features of *Dasein* [56, 57].

"Being-in-the-world" means that he/she is essentially in tune with the world and is involved with others and tools in his/her everydayness, is familiar with his/her environment, and pays attention to them. Thus, when we are immersed in the routines of our everydayness, we are in tune with the world in a homelike mood-*Stimmung*- and we encounter the world in the pre-reflection background of familiarity with the world and things [56, 57]. Thus, "being-in-the-world" can be understood as "homelike-being-in-the-world". In the following, with such an understanding of human beings, we will explain the concept of illness and health.

Unconcealment of the Homelessness- (Unheimlichkeit)

As mentioned, when we are immersed in the everyday routine of life, we are in a familiar mood with the world and our world's enduring background encompasses us. Before the pandemic, our existence was mostly sheltered in this rhythmic mood, where things made sense to us, events were predictable, coherent, and consistent and the future opened up a range of realizable possibilities. Nevertheless, the pandemic has transformed this sense of familiarity. We suddenly have found ourselves in a situation where things no longer make sense as they did in the past. Our reassuring grip on things has become slippery and unstable and as a result, feelings of anxiety and unfamiliarity have taken over our being [55].

The pandemic by disruption of everydayness has challenged our understanding of bodies, communications, and previous lifestyles in general and has revealed what has not been visible to us before. Kevin Aho calls this situation Uncanniness or "not-being-at-home"; a phrase first mentioned by Heidegger. From Heidegger's standpoint, as Dasein is thrown into the world[2] and is "being-with" Others, he/she is not completely at home, even in the moments when he/she experiences the feeling of being "at-home". Homelessness has always overshadowed our familiarity with the lived-world: this is my world, but at the same time, it is not entirely mine; I do not know it completely and cannot control it. I am thrown into the world, and I am among others; therefore, my world, despite mineness,[3] is also otherness. However, we often do not realize this homelessness and suppose our everydayness mood is

---

[2] Geworfenheit.
[3] Jemeinigkeit.

homelike [58]. A crisis like the COVID-19 pandemic reveals this hidden aspect and reminds us of "not- being- at-home".

What has made humans ill and solicitous during this pandemic and has been their reminder of homelessness could not be considered only the physical defects and disorders caused by this virus or the danger that threatens physical and mental health. However, scientific knowledge can help to explain the causal relationships of many phenomena, but it is not sufficient to understand our everyday world. Understanding this illness requires a phenomenological perspective on the illness and health and attention to one's "being-in-the-world" [59]. In other words, the starting point of the phenomenological approach in contemporary philosophy of medicine is a positivist approach to the body that cannot embrace other aspects of human experience, including social, cultural and existential; hence, it fails to perceive humanistic experiences [60]. The phenomenology of illness informs us about fundamental impacts and changes that illness has on the human's "being-in-the-world" and the conceptual fusion of these two concepts; also, it will help to understand the essence of human existence and define the concept of health based on it [59]. Although Heidegger did not address health and illness himself, Fredrik Svenaeus has used Heidegger's term "not- being- at-home" to study the phenomenology of illness. In his book "*The Hermeneutics of Medicine and the Phenomenology of Health: Steps Towards a Philosophy of Medical Practice*" [58], he says that a physical or mental disease, by affecting one's lived-body, affects other existential aspects as well and, as a result, makes a person ill and turns his/her "being-in-the-world" unhomelike [58]. It is noteworthy that in this article, we should differentiate disease as a biological dysfunction and illness as one's lived-experience of not being well. The phenomenological illness that Svenaeus envisions is neither entirely "mental" nor "physical," but rather a kind of rupture in one's "being-in-the-world". This disturbing ill mood influences the entire existence [58]. Nevertheless, this illness as "not-being-at-home" appears only when we lose health in its phenomenological sense.

Hans Georg Gadamer, in his book "*The Enigma of Health*" [61], illustrates that "Health is not a state that one introspectively feels in oneself" and describes health as "the rhythm of life" [62]. Svenaeus, who is influenced by Gadamer, also considers health as "homelike-being-in-the-world" and States: "When we are healthy, everything "flows", the mood we find ourselves in does not make itself heard or seen (…) Health is a non-apparent attunement, a rhythmic, balancing mood that supports our understanding in a homelike way without calling for our attention" [58]. When we are healthy, the danger of slipping into homelessness and therefore losing the homelike attunement is held at a distance. Health is that moment when a homelike mood overcomes an unhomelike one and encompasses our "being-in-the-world" [63]. This homelike mood certainly does not only mean feeling well, being fit, not having pain, not having signs and symptoms of a disease, etc. These are possibilities that health brings with itself rather than the true concept of health. Health is a neutral "being-in-the-world" that we do not recognize until it gets lost, so it is not understandable merely from a psychological or biological perspective [63].

If we assume health as "homelike-being-in-the-world", then illness is a disorder in our attunement with the world and a metaphor of homelessness. The illness caused by a physical or mental disease threatens our usual familiarity with the world and destroys our unified understanding of it, hence makes the world unhomelike [64]. Following this disorder in sense-making, it is certainly not the one's body that becomes alienated and enigmatic; rather, the whole ways of "being-in-the-world" are also affected. Moreover, disorderliness in meanings, routines, expectations, and connections that have changed or disappeared can provoke feelings of fear, isolation, and anxiety [64].

Illness disrupts most of the person's previous possibilities as if the person has been thrown out of everydayness, just like Heidegger's famous example of the hammer. We do not recognize the hammer while using it unless it suddenly falls to the ground, and this is the only moment in which we pay attention to it. Svenaeus has studied illness caused by changes in the lived-body. Nevertheless, the situation created in the COVID-19 pandemic goes beyond his interpretation because, in the current pandemic, the fundamental rupture in familiarity and attunement with the world is not necessarily related to the disease but also exists even without infection with this virus [62]. In other words, even those who are considered healthy based on the common health criteria may find themselves ill in this crisis. The COVID-19 pandemic has revealed "not-being-at-home" by breaking lots of our daily habits.

Unconcealment of the health is the consequence of illness; the ill person starts to ask whether and how she/he has "fallen out of things" and why things are not like what they were before [64] "The sick person," says Gadamer, "is no longer simply identical with the person he or she was before. For the sick individual 'falls out' of things, has already fallen out of their normal place in life" [61].

Policies such as increasing social distance, reducing physical contact or quarantines, etc. for controlling the spread of the virus have made people more isolated and made them feel like they have fallen out of their normal way of communication. This isolation and being far away from others has crucial impacts on people feeling "not-being-at-home" during this pandemic because, as we have already mentioned, one of the fundamental features of Dasein is "being-with" others in the shared world.

Therefore COVID-19 pandemic is a phenomenon that has altered our "being-in-the-world" and faded the vision of a predictable life, making the present horrific and frightening and the future ambiguous. In Kevin Aho's words, the human experience in the current pandemic can be called "ontological death", which means in addition to the possibility of death from a disease, we can experience death because the intelligible world that we draw on to shape our identities and retain our sense of self has lost all significance as well [55, 65]. Rupture and collapse in finding the world predictable and meaningful disable us to understand and give meaning to who we are. We can no longer think of ourselves as someone or something because we have lost the comprehensible world we identify ourselves with. On this account, death can happen numerous times throughout our finite lives and is not only an occurrence that we can physiologically go through [66].

Now, what can we do with the inability and shortages of the modern health system in perceiving existential human experiences due to the positivism of modern medicine? Undoubtedly, the purpose of asking this question is not to add an unreasonable burden on the health system, especially physicians; since medicine is inevitably intertwined with the meaning and concept of health, achieving a new understanding of health will affect medical theory and practice as well. In the following, by criticizing the ontical concept of health, we will address the reasons for the inadequacy of modern medicine in facing existential aspects of pandemics.

Reflection on Homelessness of the Modern Health System

According to previous explanations, we would realize that the pandemic has unconcealed our "not-being-at-home" condition, generating anxiety. However, the current health system strategies for confronting disease expansion and sustaining human physical health are limited to pieces of advice like wearing masks, washing hands, social distancing, etc. or attempts to achieve vaccine or proper treatment. In our opinion, these strategies are based on the health system's insufficient understanding of the concept of health; any understanding of the health concept has philosophical fundamentals and represents the way we understand and perceive the human being [67]. The World Health Organization defines health as "a state of complete physical, mental and social well-being and not merely the absence of disease or infirmity" [68]. However, this definition is based on the ontical understanding of the human, as if human consists of several components that the resultant of their function represents his/her health state [67]. It seems this classification is beneficial for scientific research but fails to approach the whole human existence.

Consequently, this sort of reductionist definition is not adequate for philosophical examinations. Besides, according to Heidegger, health is an existential condition, and we believe the ontological understanding of a human as a whole would be neglected if we conceive the human solely as a set of physical, mental, and social components [67]. WHO guidelines are based, as mentioned, on the ontical concept of health and chases for sustaining the physical and mental health of the population, but this pandemic recalled and unconcealed the embedded crisis within the current health system, which is literally the neglect of human existence as a whole. Recommend guidelines in the face of COVID- 19 pandemic, with a total gaze on physiological health, lacks a phenomenological viewpoint of human, thus fail to conceive the existential illness and the human aspects of such conditions. In this crisis, conventional response to "not-being-at-home" and the anxiety that comes with it could include two ways: (A) attempt to escape from the crises and immediate return to the previous situation or (B) facing the illness phenomenologically until achieving a more authentic and purposeful life [64].

In our opinion, usual strategies for coping with crises like the COVID-19 pandemic, instead of acknowledging our instability and existential uncertainty and critical and authentic encounter with existential anxiety, are more of an escape from existential reality and an immediate return to the previous condition. Undoubtedly this is rooted in the unwillingness of the modern health system to face and admit human vulnerability and considers the disease a terrible and pitiful condition that must be avoided in any way possible. As Arthur Frank declares, the paradigm of modern medicine is based on "restitution" that merely considers identification of the biological dysfunction, proper intervention, and eventually resituating of health as its responsibility [69]. In other words, it leads to a set of positivist expectations like there must be a proper treatment for each disease. Consequently, present endeavours to return to the pre-pandemic condition can be comprehensible this way.

During the contemporary crises, we observed how the absence of treatment or preventive guidelines challenged the worldwide health system, which is solely focused on finding a way to restrict the expansion of the virus as soon as possible. Therefore, the fact that the health system conceives its responsibility merely as resituating and reform the dysfunctioning body represents the positivist understanding of human and the concept of health and as we can realize in crisis as such, it neither acts well in resituating nor can understand the human suffering and existential illness. Because when it is impossible to resituate the disabled function, clinicians do not have much left to do, and the patient's existential suffering would remain neglected. In this pandemic, it is obvious that patients try to escape their suffering and existential anxiety caused by disease and illness [64]. It is also easy to realize that all the attention is focused on the treatable pathophysiology, which is much easier to handle for the health system in comparison with pain, existential suffering, and the possibility of imminent death. The current worldwide solution is the same as the most common response to such anxiety, inauthentic turning away from it and ignoring and escaping from the human existential reality [64].

Nevertheless, we believe existential anxiety that consequently "not-being-at-home" comes into being reveals the fragile and shaky reality of existence, that we are essentially mortal and vulnerable beings toward death. However, according to Heidegger, we rather turn to some common interpretations that seek to level down this reality instead of facing the existential reality revealed by anxiety [70]. Inattention to existential anxiety and homelessness leads to the idea that such phenomena that destroy the world's meaning could be taken under control. Although scientific and positivist understanding of death and suffering while concealing much of the lived-experience of such phenomena presents itself as the best way to measure and manage such issues. Perhaps facing illness, disease, and death in modern medicine imposes a higher sense of security and is even praised as the best way of management; therefore, adopting an approach like this would ignore the hidden meaning behind these phenomena [64].

On the other hand, facing this anxiety instead of negligence would set us free from it and give us new possibilities that used to be obscure [64]. Hence the anxiety during the COVID-19 pandemic comes with a feeling of "not-being-at-home". However, it can deprive some previous opportunities, but it can pave the way, open a new way of thinking and allow us to face the crisis more authentically. This way of thinking challenges the fundamental basis of the current health system and reveals the necessity of paying attention to forgotten human existence. Unquestionably this requires the health system and modern medicine to move beyond the biomedical paradigm and focus on human existence.

A Way Towards the Possible Condition for Homecoming

Understanding mortality and vulnerability is the offered possibility for the homeless human in the age of COVID-19. This anxiety is a reminder of our responsibility toward each other's vulnerability. This reminder surely makes us question the previously taken for granted assumption. In such a case, revival or return to health, which we understood as "being-at-home", is not possible only by protecting bodies from the taint of the virus and trying to eradicate it. Since health is a matter beyond recovery of the dysfunctional body, the body may regain its previous condition; however, equilibrium and attunement of the person with his/her world are still disrupted. On the other hand, it is possible that the feeling of illness is healed while the physical recovery has not been acquired yet [64]. In other words, the third-person view of modern medicine needs to be blended into the first-person view, i.e. the patient himself who lives the experience of the disease [60]. Absolutely the pandemic cannot be reduced to patient-physician encounters. However, there is a potential ability in this crisis that, not directly but representatively, can make humans face their vulnerability through understanding the fundamental connection of "being-in-the-world" of one another. Perhaps recognizing that we are all together in one world and sharing the same potential of suffering, love, and death can lead healthy and ill people toward each other and their way home [64]. At this point, the responsibility of health care providers, along with the prevention of physical illness, is trying to understand patients' "not-being-at-home" condition and seek to turn it into "being-at-home" or at least close to home. In companion with ill individuals, clinicians must see their everyday issues from a phenomenological perspective [58] and help them make sense of their suffering and find a way home [70].

Acknowledging and accepting existential concepts like "ontological death" helps broaden the biomedical view of suffering and illness and not reduce them to merely physical pain and discomfort. It also raises attention to ontological suffering, which goes together with the collapse of the world for the patient. However, ontological Death is a challenge facing the current health system because it lacks the proper potential to meet existential concepts. This pandemic once again reminds us that the health system needs to create such a dialogic platform that could interpret the suffering and experience of rupture in meaning and everydayness from an existential perspective and accept their vulnerability [66]. Put differently, health care providers, in addition to recognizing the physical aspects of illness and their treatment [60], must learn how to help the patient make sense of his/her suffering and to be able to re-open him/herself to his/her possibilities. Therefore, along with considering the scientific achievements, addressing the issues of disease and health requires rehumanizing the current common approach by rethinking the quality of human presence in the world (Fig. 7.3) and his/her experience of health and disease and avoiding pure attention to the physical and physiological aspects of the disease. We believe this matter will not be possible except by adopting a humanities-based approach, especially focusing on philosophizing and asking fundamental questions [71].

Fig. 7.3 Photo by Saba
Miri kermanshahi

1. John Berger wrote that "There are, in any case, occasions that defy either time or any time". Do you think that the pandemic has defied time? And has our idea of time changed?
2. Are human beings capable of learning from the past or do they continually repeat past mistakes?
3. Heidegger claims, "Questioning is the piety of thought". What do you think about the role of thinking in our lives?
4. In one of his poems, Hölderlin[4] states, "But where danger is, deliverance also grows". How could we reflect on the pandemic crisis in the same way?
5. Have you ever felt in this pandemic that you have lost your daily life?

---

[4] **Johann Christian Friedrich Hölderlin** (UK: /ˈhɜːldərliːn/, US: /ˈhʌl-/ [1]; German: [ˈfʁiːdʁɪç ˈhœl.dɐ.lin] (20 March 1770–1777 June 1843) was a German poet and philosopher. He was a key figure of German Romanticism and had a great influence on Martin Heidegger.

References

1. Eliot TS. Four quartets. London: Faber and Faber; 1941.
2. Abraham R, Ueda Y. The chaos avant-garde [Internet]. World Sci. 2001;39. Available from: https://www.worldscientific.com/worldscibooks/10.1142/4510.
3. Coronavirus: China reports 17 new cases of Sars-like mystery virus [Internet]. the Guardian. 2020 [cited 2021 June 12]. Available from: http://www.theguardian.com/world/2020/jan/19/coronavirus-china-reports-17-new-cases-of-sars-like-mystery-virus.
4. Soriano JB. Humanistic Epidemiology: love in the time of cholera, COVID-19 and other outbreaks. Eur J Epidemiol. 2020;35(4):321–4.
5. Fichte JG, Preuss P. The vocation of man. Print. 1991.
6. Weatherall DJ. Cyril Clarke and the prevention of rhesus haemolytic disease of the newborn. Br J Haematol. 2012;157(1):41–6.
7. Winston Churchill's butterfly house brought back to life. Butterflies. The Guardian.
8. Wings of desire: why the hobby of Butterfly collecting is over—it's all about conservation now. The Independent.
9. Fowles J. The collector. London: Reprint Society; 1964.
10. Nabokov theory on polyommatus blue butterflies is vindicated. The New York Times.
11. The School of Life. How to think more effectively. London: The School of Life; 2020.
12. Dante Alighieri 1265–1321. The divine comedy of Dante Alighieri: inferno, purgatory, paradise [Internet]. New York: The Union Library Association; 1935. Available from: https://search.library.wisc.edu/catalog/999855868602121.
13. Dylan B. Dignity [Internet]. Bob Dylan's greatest hits, vol. 3. 1994 [cited 2021 June 12]. Available from: https://music.apple.com/us/album/dignity/157451373?i=157451673.
14. Lovatt S. Birdsong in a time of silence. Penguin; 2021.
15. Beatles T. Blackbird [Internet]. The beatles (the white album). 1968 [cited 2021 June 12]. Available from: https://music.apple.com/us/album/blackbird/1441133180?i=1441133834.
16. Udwadia Z. India's Covid wards are like scenes from Dante's 'Inferno' [Internet]. 2021 [cited 2021 June 12]. Available from: https://www.ft.com/content/ad200d93-3247-409a-8afb-482234h4655c
17. Arundhati Roy On India's Covid catastrophe: 'we are witnessing a crime against humanity' [Internet]. The Guardian. [Cited 2021 June 12]. Available from: https://www.theguardian.com/news/2021/apr/28/crime-against-humanity-arundhati-roy-india-covid-catastrophe.
18. El Apartheid AL. De 2021: Las Vacunas [Internet]. Diario De Mallorca. 2021 [cited 2021 June 12]. Available from: https://www.diariodemallorca.es/opinion/2021/05/11/apartheid-2021-vacunas-51657766.html.
19. Surgeons and the soul - A book of words - Rudyard Kipling, book, etext [Internet]. [cited 2021 June 12]. Available from: http://www.telelib.com/authors/K/KiplingRudyard/prose/BookOfWords/surgeonssoul.html.
20. People's party wins madrid snap election but fails to get majority [Internet]. The Guardian. [Cited 2021 June 12]. Available from: https://www.theguardian.com/world/2021/may/04/voting-under-way-madrid-snap-election-dominated-covid-spain.
21. In The Words Of Nelson Mandela, To be free is not merely to cast off one's chains. Social Impact Field Seminar [Internet]. © 2021 Boston University. [cited 2021 Jun 12]. Available from: http://sites.bu.edu/socialimpactblog/2019/04/01/in-the-words-of-nelson-mandela-to-be-free-is-not-merely-to-cast-off-ones-chains/
22. Raymond C. A new path to the waterfall. London: Collins Harvill; 1990.
23. Zizek S. Pandemic! Covid-19 shakes the world. Cambridge: Polity; 2020.
24. Horton R. Offline: thinking the pandemic. Lancet. 2021;397(10276):780. https://doi.org/10.1016/S0140-6736(21)00514-6.
25. Horton R. Offline: COVID-19 is not a pandemic. Lancet. 2020;396(10255):874. https://doi.org/10.1016/S0140-6736(20)32000-6.

26. Nuwer R. Andy warhol probably never said his celebrated "fifteen minutes of fame" line [Internet]. Smithsonian Magazine. [cited 2021 June 12]. Available from: https://www.smithsonianmag.com/smart-news/andy-warhol-probably-never-said-his-celebrated-fame-line-180950456/.
27. Bauman Z. Liquid modernity. Wiley; 2013.
28. Cabane OF. The net and the butterfly. Penguin; 2017.
29. Butterfly. Origin and meaning of butterfly by online etymology dictionary [Internet]. [Cited 2021 June 12]. Available from: https://www.etymonline.com/word/butterfly.
30. Georgoulis KD. Psyche. Helios Encyclopaed Lex. 1975;24:239–46.
31. Ramón y Cajal S. Recollections of my life. Birmingham, AL: Gryphon; 1988.
32. Nabokov VV. Speak, memory: an autobiography revisited. New York, NY: Putnam; 1966.
33. Shakespeare, W. The tempest, 4th ed., Gill R, editor. London: Oxford University Press; 2001.
34. Somers J. How the coronavirus hacks the immune system [Internet]. The New Yorker. 2020 [cited 2021 June 12]. Available from: https://www.newyorker.com/magazine/2020/11/09/how-the-coronavirus-hacks-the-immune-system.
35. Shakespeare W. Hamlet, Act 3, scene 2, lines 20–21, 4th ed., Gill R, editor. London: Oxford University Press; 2001.
36. Koonin EV, Starokadomskyy P. Are viruses alive? The replicator paradigm sheds decisive light on an old but misguided question. Stud Hist Phil Biol Biomed Sci. 2016;59:125–34. https://doi.org/10.1016/j.shpsc.2016.02.016.
37. Aristotle. Politics [Πολιτικά], Book 1, ch. 2/1253a.2 [tr. Rackham (1932)].
38. Pausanias. Pausanias description of greece with an english translation by Jones WHS, Litt D, Ormerod HA, in 4 volumes. Cambridge, MA: Harvard University Press; London: William Heinemann Ltd; 1918.
39. Sands S. The interior silence. Hachette; 2021.
40. Rinpoche S. Tibetan book of living and dying. Rider. 2017;
41. Michael C. The future will have to wait [Internet]. 2006 [cited 2021 June 13]. Available from: https://longnow.org/essays/omega-glory/.
42. Stoppard T. Rosencrantz and guildenstern are dead. Faber & Faber. 2013;
43. Rovelli C. There are places in the world where rules are less important than kindness. Penguin; 2020.
44. Mukherjee S. The laws of medicine. TED Books; 2015.
45. About Jonas Salk - Salk institute for biological studies [Internet]. Salk Institute For Biological Studies. [cited 2021 June 13]. Available from: https://www.salk.edu/about/history-of-salk/jonas-salk/
46. Tesson S. The art of patience. Oneworld; 2021.
47. Berger J. What time is it? 2019.
48. Shakespeare W. As you like it by William Shakespeare - Delphi classics (illustrated). Delphi Classics; 2017.
49. U2. running to stand still [internet]. The Joshua Tree. 1987 [cited 2021 June 13]. Available from: https://music.apple.com/us/album/running-to-stand-still/1443155637?i=1443155649.
50. Mario B, Julio A, editors. Memorias de la COVID-19 [Internet]. Madrid: REDTBS; 2020. Available from: https://memoriasdelacovid19.org/libro/
51. Coope J. On the need for an ecologically dimensioned medical humanities. Med Humanit. 2021;47(1):123–7. https://doi.org/10.1136/medhum-2019-011720.
52. Berger J. A tragedy the size of a planet. The Guardian [Internet]. 2001 [cited 2021 June 13]. Available from: https://www.theguardian.com/culture/2001/may/28/artsfeatures.globalisation.
53. Favero A. Te quiero [Internet]. Mario Benedetti- Alberto Favero. 2002 [cited 2021 June 13]. Available from: https://music.apple.com/us/album/te-quiero/275336857?i=275336938.
54. Cookson C. WHO and global leaders could have averted Covid catastrophe, say experts [Internet]. 2021 [cited 2021 June 13]. Available from: https://www.ft.com/content/fb698a43-0f97-4142-9ff5-15dde66ffae0.
55. Aho K. The uncanny in the time of pandemics: heideggerian reflections on the coronavirus. Gatherings: The Heidegger Circle Annual. 2020;10:1–19.
56. Inwood M. A Heidegger dictionary. 1999.

57. Inwood M. A very Short Introduction to Heidegger. Oxford: Basil Blackwell Publishers; 1997.
58. Svenaeus F. The hermeneutics of medicine and the phenomenology of health: steps towards a philosophy of medical practice. Springer; 2000.
59. Ferguson T. I am not myself: understanding illness as unhomelike being-in-the-world. AMA J Ethics. 2012;14(4):314–7.
60. Svenaeus F, editor. A defense of the phenomenological account of health and illness. J Med Philos Forum Bioeth Philos Med. Oxford University Press; 2019.
61. Gadamer H-G, editor. The enigma of health: the art of healing in a scientific age. Wiley; 2018.
62. Gadamer H-G, Weinsheimer J, Marshall DG. EPZ truth and method. Bloomsbury Publishing; 2004.
63. Ahlzén R. Illness as unhomelike being-in-the-world? Phenomenology and medical practice. Med Health Care Philos. 2011;14(3):323–31.
64. Aho K. Kevin Aho: existential medicine: essays on health and illness. 2018.
65. Aho K. Why Heidegger is not an existentialist: interpreting authenticity and historicity in Being and Time. Florida Philosophical Review. 2003;3(2):5–22.
66. Aho KA. Heidegger, ontological death, and the healing professions. Med Health Care Philos. 2016;19(1):55–63.
67. Sarvimäki A. Well-being as being well—A Heideggerian look at well-being. Int J Qual Stud Health Well Being. 2006;1(1):4–10.
68. Available from: https://www.who.int/about/who-we-are/constitution.
69. Frank AW. The wounded storyteller: body, illness, and ethics. University of Chicago Press; 2013.
70. Svenaeus F. Das unheimliche–towards a phenomenology of illness. Med Health Care Philos. 2000;3(1):3–16.
71. Stempsey WE. Medical humanities and philosophy: is the universe expanding or contracting? Med Health Care Philos. 2007;10(4):373–83.

Heroes, or Rather Not. The Healthcare Professionals' Year of the Pandemic

Paola Chesi

> *"E il mio maestro mi insegnò com'è difficile trovare l'alba dentro l'imbrunire."*
>
> *"And my master taught me how hard it is to find the sunrise down in the twilight."*
>
> Franco Battiato

Being Healthcare Providers Before the Pandemic

As researchers dedicated to the analysis of care pathways, in the Fondazione ISTUD we have always focused our attention on healthcare professionals' wellbeing, the main resource for any healthcare system.

From our surveys and courses carried out throughout Italy before the pandemic, we could depict a transverse scenario of burnout and compassion fatigue among care providers, mainly caused by organizational issues: excessive workload, administrative duties, loss of autonomy, managerial objectives to reach, inadequate support staff, the cutting of hospital beds, and the reduction of care services. To give an example of the cuts in the Italian, in 2012 the Intensive Care Units had 12.5 beds per 100,000 inhabitants, compared to the 29.2 of Germany.

On the other hand, in the middle of a digital revolution, the establishment of new care relationships had started; from a more paternalist approach—"all that my doctor says is good for me"—the patients had finally become more involved in the

P. Chesi (✉)
Healthcare and Wellbeing Area at ISTUD, Milan, Italy
e-mail: pchesi@istud.it

M. G. Marini, J. McFarland (eds.), *Health Humanities for Quality of Care in Times of COVID -19*, New Paradigms in Healthcare,
https://doi.org/10.1007/978-3-030-93359-3_8

choices for their own treatment. This positive and important goal coincided with other phenomena, such as the proliferation of misinformation, and the expectation of higher standards of health; even immortality. Thus, legal issues from unsatisfied patients became part of the health professionals' working days, and fear of medical litigations was the main factor influencing the adoption of defensive medicine behaviours [1].

From some differences, this scenario was not an Italian peculiarity. Recent scientific bibliography has published many papers on the effects of burnout and emotional distress on healthcare professionals in challenging care settings such as neurology—"the neurologists' crisis"—intensive care units, oncology, and palliative care [2–5].

More, medically-related professions were at high suicide risk in different countries, particularly women-physicians and specialists from anaesthesiology, psychiatry, general practice, and surgery [6].

I remember a significant episode when I attended the Congress of the Italian Society of Neurology in Rome, at the end of 2018, to present a research on compassion fatigue among neurological teams caring for Multiple Sclerosis [7]. The topic of the survey was very different from the others in the same session; at the end of my presentation, the president of the commission commented: "Finally we are talking about us!". Several scientific societies began to realize that there was a need to face the decreased health professionals' wellbeing.

Then, the Pandemic Came

From the first cases of SARS CoV-2 in Italy—March 2020—a new sudden scenario burst into the whole healthcare system. For the healthcare professionals directly involved in the Covid cures, the previous mechanism was replaced with a new and brutal emergency. For those who were not directly involved, everything or almost everything stopped. For both categories, it was a trauma. In the first case, the trauma was represented by hospitals full of patients, the impossibility to cure many of them, the abnormal number of deaths, the lack of physical contact, the overall lack of humanity caused by the emergency, and by the risks of contagions; and much more, the ethical dilemma, the isolation from the own families, the endless work shifts. In the second case, the impossibility to continue to take care of their own patients, the gradual de-structuring of the care services—spaces and staff occupied by the cure for Covid—and the necessity to find new ways to cure were the main traumatic conditions.

From March to April 2020, we carried out a narrative survey addressed to the general population on living through the lockdown [8]. Thirty-two percent of the participants were health care providers, mainly coming from our multidisciplinary network. We collected testimonies from those in the first line, living the dualism between the total dedication to their own profession—"*I think that it's correct to respect the code of professional conduct, including the risk of contagion while doing the own job*"; "*I feel ready to die. I am a physician, and this is my identity; if I die,*

I die for who I am"—and the helplessness and suffering in front of so many deaths: *"I have to communicate to family members who don't know me that their loved one has passed away, hearing cries of people whom I have never met"*; *"During the night, I think about all the persons who have passed away without their loved ones; it is terrible, my heart is lacerated"*; *"I feel helpless. I couldn't do anything to help the patients who passed away"*.

Among the many emerging elements, they described the new dressing process—*"Mask, visor, white coat, gloves, am I dressed correctly? Is the procedure correct? Sometimes I feel like an idiot talking to myself but this helps me to maintain my concentration"*; *"With this dressing, we are all similar, it's difficult to recognize each other; I think how difficult it must be for the patients to find a point of reference"*—and the concern coming back to their families—*"I go to work for my spirit of service but I cannot hide my fault, for me and my family"*; *"My sons avoid hugging me, and if I sneeze, they ask me to swab"*; *"I sleep separately from my wife"*; *"I tell little about my work shift to my wife, and she asks me little"*.

From these narratives we also began to comprehend their emotions, between apprehension—*"The messages from my colleagues are alarming; I turned off the bleeper, and I read them just once a day because they increase my anxiety"*; *"I feel like I have worked in a place made of paper with a lighter in my hand"*—but also trust, and the re-discovered motivation—*"I rediscovered the meaning of being a physician: every time I dress the so uncomfortable medical devices, I feel aware that I can find inside me what I can be both as a human being, and health professional"*- and sense of belonging within the care teams. All of them were acting with the same purpose, together living through something that probably would never be shared with others, not even family members: *"We could be as we are. The circumstances forced us to overcome the differences in roles and sympathy. We supported each other. We cured our colleagues with the awareness that we could have been in their shoes. Some of them passed away. Some of us lost family members. Our faces are marked by tiredness and masks, but I think we are beautiful, even with the bags under our eyes, and pallor"*.

We also collected narratives from healthcare professionals who were not on the front line and lived through the sudden changes of their job, the distance from their patients, and the concern for their health. From a gynaecologist: *"I asked a patient: why did you come to visit if this is not an emergency? And she answered: because I trust you, if you come to work, I can come to visit. From that moment, I understood that my duty was staying at home and let my patients at home"*. From a family doctor: *"My clinic is empty. I spend all day on the phone, in front of the computer, in contact via digital platforms and Whatsapp. I miss the creative and light energy. I miss the direct contact with people. I am used to catching information from gestures, posture, gazes, voices… all these things are not possible at this moment. I feel isolated even if I am communicating all day long"*. From a paediatrician: *"I am thinking about my patients with autism, with their habits, locked up with their parents, not used to staying together all this time"*. From a psychiatrist: *"I am worried that the virus could come into my patients' residential community; the place is tight and most of them are physically fragile, it would be a disaster"*.

For all the other people, they became "heroes", and the media contributed to spread this vocabulary. Photos of exhausted nurses and physicians with faces marked by the masks or with their names written on the back of the white coat circulated everywhere and became part of the media narrative of this first chapter of the pandemic. The healthcare professionals did not like this definition since they felt they were just doing their job, and would have appreciated being able to perform it into a more prepared healthcare system, and with an earlier professional recognition: *"I feel I am everything, except a hero. I don't like this expression, and this doesn't comfort me"*; *"I would like healthcare workers to be recognized as professionals rather than heroes"*.

Immediately after the lockdown—May 2020—when the numbers of contagions and deaths started to decrease, a group of healthcare professionals who were attending our Master in Applied Narrative Medicine and who experienced the cure for Covid in Lombardy, the most hit area in Italy, decided to collect narratives from their colleagues [9]. Only half of the invited people agreed to tell their story. Many of those who did not accept the invitation, said that going back to those days was too painful and that they were still emotionally overwhelmed. Below are some of the main elements that emerged:

- The crying: almost all the healthcare professionals wrote about having cried during the work activities—*"Hiding the tears under the mask"*; *"I've earned reserves of tears that I've never been able to cry at all"*.
- The importance of the eyes; because of the use of Personal Protective Equipment, the eyes remained the only means at disposal to relate with patients and colleagues—*"It was necessary to transmit humanity through the eyes"*; *"How many eyes I have imprinted in my head that look at me"*.
- The patients' loneliness: forced to be distanced from their loved ones, thousands of people lived their illness deprived of contact with their family members. The health professionals tried to replace the distant affections, preserving the memories of the many stories—*"Sick people died alone, without receiving visits from relatives, with their hands in the glove of an unknown operator who gave them the last caress"*.

The authors concluded their work with this reflection: *"the narrative medicine tools gave us the possibility to read and give voice to a complex professional experience. We wish the memory of these months could be useful to redesign better care services, with greater attention to the health professionals' work quality …not to forget"*.

Today, Not Heroes Anymore

After the first phase of the pandemic, when in Italy part of the population thought that everything had been overcome and the summer holidays came, healthcare professionals remained with their trauma, their tiredness, and their awareness that it was not finished yet. When the first signals of the so-called "second wave" came, the society was not united anymore to face the common enemy but divided amongst the different ideas, needs, and new consequent emergencies. For part of the population, they were not "heroes" anymore, rather "alarmists", in competition with the economic needs. From the media perspective, they were catapulted into a general scenario of "everyone against everyone" made of opposing opinions, till the extreme diffusion of the idea of a "general conspiracy". The characteristics of this virus and the several unknown elements did not help. In autumn 2020, within this climate of increasing anger and impatience, the healthcare providers, who were not recovered yet from the first phase, found themselves in the second.

In November 2020, we carried out an online course named "The languages of care", focused on the role of arts applied to healthcare. We were in the middle of the emergency, almost all of Italy was in a red zone, and I doubted that part of the participants, especially the health workers from the intensive care units, would have been able to attend the course. But they "fought like lions" to obtain the permissions to participate, and wrote me sentences such as: "*Please, don't cancel the course, don't leave us*"; "*It doesn't seem real that for three days I won't talk about Covid, they will be the first carefree days of the year*"; "*I am waiting for the course with a big desire of beauty to recover from this chaos*". They needed a space in which to express themselves, share their emotions, distract just for a while, breath, and be cared for, instead of care for. Through the course, we supplied them with tools from the arts to help them recover from their trauma and bring them wellbeing. We directly experimented with the positive effects of paintings, music, games, colours, poems, and dance (yes, we danced!). Along the 3 days course, the faces became more relaxed and smiling, and at the end, we received the following feedback: "*I feel a vital euphoria. I wish you could organize a sequel to this course*" [and we will do it]; "*Thank you for these days, I felt beauty and serenity*". We also aimed to measure the effects of the use of arts through the Maslach Test of Burn Out, distributed among the participants both before and after the course. In all the cases, the value of burnout reduced by 2%; in particular, frustration, tension, and tiredness decreased, whereas self-esteem increased.

The distinction between the second and the third wave—occurring in February/March 2021—was less perceived, especially by the healthcare professionals; some of them stated that the third phase came when the second one was not yet even finished, as in a continuum timeline. In a sort of a circular paradox, Lombardy, the most hit area at the beginning of the pandemic, lived another strong emergency. But while one year before the entire community had joined together into a common and sympathetic effort, in this third phase the climate was different: disenchantment, delusion, anger for the management of the emergency and of the vaccination campaign, mixed with tiredness and frustration.

We have been continuing to collect healthcare professionals' narratives from different areas and contexts of care in Italy. The dualism between apprehension and trust is still present. But during this year, most of them did not have a space to share emotions and reflections within the care teams, since the organizational issues—work shifts, the opening and closing of the wards, protection measures—continued to be predominant. If already before the pandemic we had collected requests from healthcare professionals for psychological support, now this need is even stronger, as shown by the most recent narratives: *"It would be helpful having psychological support within our ward, not only for patients but also for us, to preserve our wellbeing. In the previous critical phases, we invested many of our resources, now we risk the emotional breakdown"*; *"The general mood of my care team has been so heavily decreased another time. By now, fatigue is insurmountable"*.

Resources for Tomorrow, to Bring Wellbeing

By March 2021, one year after the beginning of the pandemic, in Italy, 340 doctors and 81 nurses have died from Covid [10, 11]; the COVID-19–related deaths include 2 nurses who committed suicide due to unsustainable pressure at work [12]. More than 3.000 healthcare workers have been reported worldwide, including other suicide cases, but numbers continue to increase [13–15].

Recent research findings confirm that many healthcare workers are experiencing symptoms of depression, anxiety, insomnia, and distress, especially those working directly with COVID-19 patients [16, 17].

Several organizations have issued guidelines with recommendations on how to protect healthcare workers' wellbeing, including psychological support, self-monitoring behaviours—such as maintaining regular sleeping and feeding habits, avoiding smoke, alcohol, and the excess of information and communication on socials—and peer-support. In some healthcare facilities, staff have access to helplines, online therapies, counselling, group support sessions, and mindfulness; all these are useful strategies to support resiliency and mental wellbeing. Nevertheless, most of the guidelines remained on a general level, and the first evidence of the concrete benefits reveals the difficulty of carrying them out in specific care contexts [18]. More, the level of trauma and the lack of experience in preserving the own wellbeing—because this kind of care was mostly neglected already before the pandemic—require practical, fast, and inexpensive support tools, together with structured support programs.

As a research and educational centre that has been working on narrative medicine for a long time, we believe that narrative is even more crucial now to welcome the need to express ourselves and our alternating emotions. Coming back to the collection of narratives on the lockdown experience, we tested the immediate effects of the writing activity using the analysis of emotions; we invited the participants to indicate the three main emotions felt before and immediately after the narrative. While at the beginning the most recurrent emotions were apprehension, pensiveness, and acceptance, later trust replaced pensiveness, and serenity appeared, although the persistence of apprehension, as stated by 67.8% of the respondents: *"writing is relaxing and gratifying"*; *"it was useful to extern what I have been keeping for days"*; *"it was an experience of liberation and abandon of negative emotions"*. This confirmed the positive role of the inner reflection offered by the writing process. More, narrative medicine develops empathy, and the attitude to the listening and sharing of perspectives, which are crucial tools for fortifying resilience and inner motivation [19, 20].

On the other hand, we saw how the difficulty of writing about trauma and pain are still intense. Many healthcare providers invited to narrate their experience with Covid-19 clearly said that they were not able to write and recall painful episodes. How to help them to overcome this block? How to supply helpful tools to those who cannot express themselves through words? In this case, the use of arts can represent the answer and the strategy to allow the expression of emotions. In 2019, WHO published the Health Evidence Network synthesis report named "The role of the arts in improving health and well-being in the WHO European Region"; the document contains demonstrations of how arts interventions can improve health and wellbeing, contribute to the prevention of a variety of mental and physical illnesses, and support in the treatment or management of acute and chronic conditions. More, arts interventions are presented as low-risk, highly cost-effective, integrated, and holistic treatment options for complex health challenges. For these reasons, WHO invited the EU countries to develop integrative programs between the arts and health and social care [21].

Together with Stephen Legari, art therapist from the Montreal Museum of Fine Arts—the first one to experience medical prescription at the museum [22]—in Fondazione ISTUD we have been experimenting with how to combine the use of famous pictures from Van Gogh and narratives with the healthcare professionals who attend our courses on narrative medicine, from 2019 to now; 165 care providers. In this activity, the participants contemplated one Vincent Van Gogh painting— more frequently, "The Starry Night"—and then wrote down their impressions, emotions, reflections, details without necessarily considering the author's context of life or biography. A basic narrative plot stimulated the reflections of the participants: *I feel .... I think............ I want to do.............* In the last part of the practice, they identified a place where they could be in the painting and invent their own story. In all the contexts in which we proposed this exercise, the healthcare professionals expressed at first a moment of surprise, and in some cases, uncertainty: "Why do I have to look at a familiar image?", "What do I have to write?". Then, silence came. The silence of the observation, of the personal dialogue of each one with the image. Gradually, words came, flowing onto the paper. We never requested written reflections, since they were the expression of an intimate moment, but afterwards, we observed a strong willingness to share the experience. Independently from the kind of emerged emotions, for the healthcare professionals the intervention represented a means of letting them share emotions within a contained experience through the mirroring capacity of the image, and, at the end of the practices, most of the participants stated that they felt better and "happier", evaluating the activity as amusing and as a distraction from their daily job. From this low-cost and low-time consuming simple intervention, it is possible to supply healthcare professionals with a tool to overcome possible blocks, let emotions flow, welcome them and, in the end, recognize them.

We can also give testimony to a successful Italian experience at the Humanitas Gavazzeni Hospital, in Bergamo, one of the most badly affected cities from

Covid-19 in Italy. In 2019, the management of the healthcare facility decided to establish a partnership with Accademia Carrara, a beautiful gallery in which famous pictures from the fifteenth to the nineteenth century are exhibited. Gradually, reprinted images from the gallery were placed within the hospital, from the entrance to the bar to specific wards [23]. Since the beginning of the pandemic, Gavazzeni has become a Covid hospital, because of the strong impact of the virus on the population of the city. After this first traumatic wave, we asked a multidisciplinary group of healthcare professionals to explain the possible effects of the paintings on the walls; "beauty" was the most recurrent word, and "serenity" was the most cited emotion: *"the images move me, I feel serenity, safety, relax"*; *"They are relaxing, I recover physical and mental energies, I can throw out negative emotions"*; *"I enter into the facility with my problems, and looking at the images I feel less pain"*; *"It is a useful moment of detachment to recharge energy"*; *"I receive a sense of warmth"*.

These are first and local experiences, and we continue to analyse the effects of the use of arts applied to healthcare, but the potentialities and possible benefits for the healthcare professionals' wellbeing are already clear to us.

I will conclude this session on art by telling a short story from an anaesthesiologist in a Covid ward: *"Yesterday at the intensive care unit, it was in the late afternoon, after a 12-hour shift, I decided to put on some music, one of those 60/70/80s compilations available on YouTube. Some patients with tracheotomy started to move their arms following the music, it was like a dance. Before I went away, one of them said to me "thank you, I spent a beautiful afternoon" …Amazing!"*. A few days earlier, I had received a mail from the same person in which she wrote to me that at that moment she was not able to write about her experience of caring for Covid since the pain was too strong. With this simple and unstructured action based on music, this physician was able to find a resource not only for herself but also for her patients, and after a few days, we received her narrative.

Why Writing About the Healthcare Professionals' Experiences?

With this partial overview, I have only aimed to show how many resources, more or less structured, are at the disposal to take care of healthcare professionals' wellbeing. Probably a uniform tool effective in every context and suitable for everyone does not exist. What is important is that protecting the healthcare professionals' wellbeing must become a priority for policymakers and health managers. The most recent scientific bibliography contains many invocations to develop strategies for monitoring the mental health of healthcare workers, immediately and after the end of this emergency [24].

I am not a clinician, but my role as researcher and tutor in narrative medicine gives me the privilege to foster several health workers in their projects with the use of narrative, applied to themselves or their context of care; with some of them, the relationship becomes so close that our exchanges of mail represents a sort of "informal diary" of their professional living. I also had this privilege during the pandemic; and even though tiredness, lack of time, and sometimes lack of words was

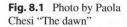

Fig. 8.1 Photo by Paola
Chesi "The dawn"

prevalent, some physicians, nurses, and others constantly wrote to me, or to the whole Healthcare and Wellbeing group of ISTUD, to share their feelings, experiences, significative episodes, or thoughts. I like to consider myself as the custodian of these "diaries", and, as a custodian, I feel a duty to welcome their requests for attention.

I am also a sister of a physician who has lived through the complete change of his profession, being moved from his usual ward, with his usual patients, to Covid care. As a psychiatrist, he was used to rehabilitating patients, and not to seeing so many deaths. He was used to having solutions, and not to fighting all night long to keep a person alive. I know that he has the inner resources to overcome all this and find a new and enriched version of the physician he will become tomorrow. But I am wondering whether health managers and policymakers are aware of this, of this turmoil, and, above all, whether will they remember the lesson learned from this pandemic.

Telling their stories is one of the contributions that I can bring to learn how to take care of our healthcare systems' pillars, and never ever forget (Fig. 8.1).

References

1. Pellino IM, Pellino G. Consequences of defensive medicine, second victims, and clinical-judicial syndrome on surgeons' medical practice and on health service. Updates in surgery. 2015; 67(4):331–7. Cunningham W, Wilson H. complaints, shame and defensive medicine. BMJ Quality Safety. 2011;20(5):449–52.
2. Busis NA, Shanafelt TD, Keran CM, et al. Burnout, career satisfaction, and Well-being among US neurologists in 2016. Neurology. 2017;88(8):797–808.
3. Henrich NJ, Dodek PM, Gladstone E, Alden L, Keenan SP, Reynolds S, et al. Consequences of moral distress in the intensive care unit: a qualitative study. Am J Crit Care. 2017;26(4) https://doi.org/10.4037/ajcc2017786.
4. Eelen S, Bauwens S, Baillon C, Distelmans W, Jacobs E, Verzelen A. The prevalence of burnout among oncology professionals: oncologists are at risk of developing burnout. Psycho-Oncology. 2014;23(12):1415–22. https://doi.org/10.1002/pon.3579.

5. Maffoni M, Argentero P, Giorgi I, Hynes J, Giardini A. Healthcare professionals' moral distress in adult palliative care: a systematic review. BMJ Support Palliat Care. 2019;9(3):245–54.
6. Dutheil F, Aubert C, Pereira B, Dambrun M, Moustafa F, Mermillod M, et al. Suicide among physicians and health-care workers: a systematic review and meta-analysis. SSRN Electron J. 2019; https://doi.org/10.1371/journal.pone.0226361.
7. Chesi P, Marini MG, Mancardi GL, Patti F, Alivernini L, Bisecco A, et al. Listening to the neurological teams for multiple sclerosis: the SMART project. Neurol Sci. 2020;41(8):2231–40. https://doi.org/10.1007/s10072-020-04301-z.
8. https://www.medicinanarrativa.eu/wp-content/uploads/Narrarsi-ai-tempi-del-COVID-1.pdf.
9. https://www.medicinanarrativa.eu/narrative-medicine-and-the-covid-19-emergency-health-professionals-as-protagonists-of-a-story-of-suffering.
10. https://portale.fnomceo.it/elenco-dei-medici-caduti-nel-corso-dellepidemia-di-covid-19/.
11. https://www.assocarenews.it/infermieri/coronavirus-ecco-tutti-gli-infermieri-deceduti-per-covid-19.
12. National Federation of Professional Nursing Orders (FNOPI) website [in Italian]. https://www.fnopi.it/. Accessed 17 Apr 2020.
13. https://www.medscape.com/viewarticle/927976.
14. Mortier P, Vilagut G, Ferrer M, Serra C, Molina JD, López-Fresneña N, et al. Thirty-day suicidal thoughts and behaviors among hospital workers during the first wave of the Spain COVID-19 outbreak. Depress Anxiety. 2021;38(5):528–44. https://doi.org/10.1002/da.23129.
15. Board PE. Health care suicides: another tragic toll of the coronavirus pandemic [Internet]. New York Post. New York Post; 2020 [cited 2021Jul3]. Available from: https://nypost.com/2020/04/27/health-care-suicides-another-tragic-toll-of-coronavirus-pandemic/.
16. Lai J, Ma S, Wang Y, Cai Z, Hu J, Wei N, et al. Factors associated with mental health outcomes among health care workers exposed to coronavirus disease 2019. JAMA Netw Open. 2020;3(3):e203976.
17. Shreffler J, Huecker M, Petrey J. The impact of COVID-19 on healthcare worker wellness: a scoping review. Western J Emerg Med. 2020;21(5) https://doi.org/10.5811/westjem.2020.7.48684.
18. Vera San Juan N, Aceituno D, Djellouli N, Sumray K, Regenold N, Syversen A, et al. Mental health and Well-being of healthcare workers during the COVID-19 pandemic in the UK: contrasting guidelines with experiences in practice. BJPsych Open. 2020;7(1) https://doi.org/10.1192/bjo.2020.148.
19. Wald HS, Haramati A, Bachner YG, Urkin J. Promoting resiliency for interprofessional faculty and senior medical students: outcomes of a workshop using mind-body medicine and interactive reflective writing. Med Teach. 2016;38(5):525–8. https://doi.org/10.3109/0142159X.2016.1150980.
20. Winkel AF, Feldman N, Moss H, Jakalow H, Simon J, Blank S. Narrative medicine workshops for obstetrics and gynecology residents and association with burnout measures. Obstet Gynecol. 2016;128(1) https://doi.org/10.1097/AOG.0000000000001619.
21. https://apps.who.int/iris/bitstream/handle/10665/329834/9789289054553-eng.pdf?fbclid=IwAR05DI_NvOMK8WTM7P6CTL1c-2Jw4ip7fVUXRzbyMOa52TaF1K5iEwsgLs8
22. https://www.mbam.qc.ca/en/art-therapy-sessions/, https://www.mbam.qc.ca/en/news/museum-prescriptions/.
23. https://www.lacarrarainhumanitas.it/.
24. Chirico F, Nucera G, Magnavita N. Protecting the mental health of healthcare workers during the COVID-19 emergency. BJPsych. International. 2020;18(1) https://doi.org/10.1192/bji.2020.39.

So Far, So Near: Telemedicine Experiences During the COVID-19 Pandemic

Fabrizio Gervasoni

> *"My view…is that the ultimate destination of all nursing is the nursing of the sick in their own homes"*
>
> *Florence Nightingale, 1867*

Florence Nightingale [1], the inspirer of modern nursing sciences, on Easter Sunday 1867, foreshadowed the desirability of taking care of the patient at home. Home care, nursing or physiotherapy, already represent a valuable assistance opportunity for patients who cannot be transported or, for some reason, are unable to reach hospital facilities for outpatient care. Home care nurses and physiotherapists continued to assist patients at their home even during the pandemic period. In doing so, however, they exposed themselves and their patients to the risk of becoming infected with SARS-CoV-2. For this reason, many health professionals were affected by COVID-19 due to exposure to the virus during home care. Therefore, hospitals and healthcare professionals had to develop new remote contact methods for patients confined to their homes. Workers in the health sector therefore organized and enhanced telemedicine and telemonitoring services.

The COVID-19 pandemic made this opportunity happen globally and implied a rapid acceleration of numerous processes, particularly technological innovation. Indeed, the health teams had to implement and strengthen telemedicine, teleconsultation, teleassistance and telerehabilitation systems. However, clinical experiences have

F. Gervasoni (✉)
ASST Fatebenefratelli Sacco - "Luigi Sacco" Hospital, Milan, Italy
e-mail: fabrizio.gervasoni@asst-fbf-sacco.it

M. G. Marini, J. McFarland (eds.), *Health Humanities for Quality of Care in Times of COVID -19*, New Paradigms in Healthcare,
https://doi.org/10.1007/978-3-030-93359-3_9

remained anecdotal and disconnected in the absence of fundamental international guidelines of good clinical practice regarding services provided in Digital Medicine.

The Recipients of Telemedicine Services

In 2020, starting from the start of the pandemic emergency, doctors and health service providers had to adopt reference protocols to prepare and provide Digital Medicine services. Researchers studied previous clinical experiences and reference models to identify the main operational problems, proposing feasible solutions supported by scientific evidence. Health service providers had to set up services that could implement new technologies in the diagnostic and therapeutic journey of patients while remaining compatible at the same time with daily clinical practice within the healthcare organization.

Healthcare professionals started to experiment telemedicine since the intensive care wards for COVID were created. Thanks to remote communication tools, tablets, and smartphones, patients infected by the SARS-CoV-2 virus reduced the detachment imposed by physical distancing and the required isolation especially during the acute phases of the disease. Patients were thus able to be in touch with family members, even during hospitalization in intensive care. Thanks to video calls with the family members, it was possible to reduce the dramatic psychological impact of physical isolation and confinement necessary for infected patients. In some virtual hospitals, doctors, nurses, and physiotherapists used technological devices (such as smartphones and tablets) to remain constantly connected with patients, reducing the risk of exposure to the virus and limit the consumption of personal protective equipment. Some rehabilitation teams have proposed remote rehabilitation protocols for hospitalized patients to limit the disabling consequences of immobility and bed rest.

Upon discharge from these virtual hospitals, some patients were provided with telemonitoring and telerehabilitation devices through which ongoing contact with the hospitalization facilities and the places of care was guaranteed. Thanks to this approach, doctors, nurses and physiotherapists were able to get virtual access to the homes of COVID patients so that a continuous follow up could be guaranteed.

In Italy, the *Istituto Superiore di Sanità* (the highest national institution on health issues) produced a document entitled "*Report of the Istituto Superiore di Sanità COVID-19 n° 12/20 - Interim indications for telemedicine assistance services during the COVID-19 health emergency*" [2]. In this document, the Italian health authorities clarified the main goals for telemedicine within the emergency COVID-19 context. The aim is to provide clinical support and care to all patients in isolation, both for quarantine resulting from clinical needs (such as acute infection by SARS-CoV-2) and the regulations in force for physical distancing.

The Report indicates the types of subjects to be targeted for telemedicine services:

- COVID-19 patients for monitoring, care and rehabilitation treatments.
- Patients exposed to close contact with SARS-CoV-2 infected individuals.
- Fragile patients with diseases, not COVID-related, who require continuity of care [2].

Is It Possible to Remain "Clinical", Even at a "Distance"?

Healthcare professionals face a new challenge: offering the usual healthcare services in a completely new, unfamiliar context, devoid of the physical proximity and human contact which are normally part of medical care. For this reason, as medical doctors, we asked ourselves if it was possible to remain "clinical" [3] (*i.e.* from Ancient Greek κλινικός *"klinikós"*, the etymology of which originates from κλίνη *"klínē"* "bed"; "close to the beds of patients"), in a context of physical distancing such as the one imposed by tele-medicine (τῆλε – tele, "far away").

Is it possible to remain empathic and close to the patient even during a remote medical visit? What language should be used to convey confidence, competence and understanding to the patient?

As a matter of fact, the very essence of the "clinician" is intrinsic in the proximity of the doctor to the patient's "bed"; while telemedicine is, by definition, "from a distance". These two aspects may seem to be incompatible. For this reason, those who offer telemedicine services should:

- Identify adequate technological equipment to conduct the visit in compliance with current legislation on privacy and personal data protection regulations.
- Develop checklists for recommending the most appropriate behaviour during the online visit, offering instructions on collecting informed consent, conducting the physical examination, and producing the clinical report (see **Practice time** at the end of this chapter).
- Consider using a new way of communication with patients, in order to allow an accurate anamnestic collection and a detailed recording of their symptoms.
- Develop a new clinical semeiology to identify the patient's signs and symptoms even in a virtual assessment context, without the possibility of direct person contact with the patient.

A remote visit should not be considered as a typical clinical evaluation, with the simple difference that is managed virtually. As we will discuss later in this chapter, this new way of providing the medical act has substantial limitations, of which the clinician must be aware, in order to minimize potential methodological errors. The report from the Italian health authorities clarifies the definition of telemedicine, describing it as a health service using computer software for the transmission of audio and video information and for the remote sharing of health documents and clinical reports. Such data exchanges between doctors and patients (and vice versa) should comply with all IT security and privacy, and personal data protection international regulations.

The *Italian Ministry of Health* [4], however, even prior to the pandemic emergency, had stated that telemedicine does not replace face to face health acts but rather integrates them. Digital Medicine represents an opportunity to quickly ensure continuity of treatment or screening for patients without being obliged for them to access hospitals, but cannot fully replace in person visits.

The Limits of Telemedicine and Remote Visits

While telemedicine represents a vital opportunity to ensure continuity of care through the provision of treatment for a larger number of patients, it also presents some risks. Clinicians need to be aware of these hazards, in order to minimize them during their remote visits. For example, in the treatment of musculoskeletal problems, telemedicine has significant limitations, both with regards to the evaluation and the rehabilitation treatment.

A telemedicine visit does not allow one:

- to perform passive mobilization of a limb;
- to evaluate the spasm of a muscle group through the passive mobilization of the anatomical segment;
- to perform provocation tests of anatomical structures;
- to palpate soft tissues, such as muscles, tendons, or ligaments, to assess their integrity or tenderness;
- to directly compress a skeletal structure in order to assess local tenderness;
- to quantify – by mobilizing it - the stability of a joint (for example of an ankle after a sprained trauma);
- to assess the strength of a muscle group, with the exception of asking patients to lift objects of known weight (for example, a bottle of water);
- to quantify the strength of some segmental movements, such as finger extensions, thumb oppositions or dorsiflexion of the ankle joint;
- to directly evaluate the strength of the handgrip required to conduct of *Activities of Daily Living* (A.D.L).

These difficulties in medical examinations already led to some clinical complications, as documented by published evidence. For example, *Surgical Neurology International* published a case report regarding two telemedicine consultants who did not identify a foot drop [5]. The article highlights the difficulty of the two medical specialists assessing the dorsiflexion of the foot in a senior subject with low back pain and severe radiculopathy. Surgery should have been considered for the patient described in the case report, but the limitations of the remote assessment delayed access to appropriate care. This case report is a tangible confirmation of the intrinsic limitations of telemedicine, which can only complement clinical evaluation, but never entirely replace it.

The Evaluation of the Patient in an Ecological Context

A physician offering online visits may find it challenging to observe the patient in a spontaneous and natural context. For example, in a virtual examination environment, it is not possible to assess the attitude of a patient when the subject does not feel observed by the doctor. A typical example is how the patient would approach the clinic prior to the visit: if using crutches or other aids, the subject would require the assistance of caregivers or would show an abnormal pace. During a telemedicine visit, the physician cannot observe some of the non-verbal attitudes of the patient or the informal interactions between the subject and caregivers. Physicians also find it difficult to observe the relational dynamics between patients and caregivers during the undressing or dressing at the end of a visit.

All these elements, collected informally during the medical examination, are often relevant to quantify the disability level of the patient, in order to set up an appropriate diagnostic and therapeutic plan.

In this context, however, telemedicine has the advantage of evaluating the patient within his residential context. Thanks to the online visit, the healthcare worker can observe the spaces available to the patient (for example, for the prescription of a wheelchair or mobility aids), can detect any potential hazard elements within their home or observe patients and their movements within an ecological and familiar context.

Telemedicine for Screening and Monitoring

Telemedicine (particularly through telemonitoring) is useful to perform an initial rapid screening of multiple patients and is functional to identify subjects who should have an on-site assessment at the hospital. In these cases, digital medicine makes it possible to identify some red flags which may be helpful in contextualizing the patient within a more in-depth clinical or diagnostic framework [5].

The physician can integrate telemedicine with a face to face approach, in order to monitor the patient during the following weeks or months. Thanks to such follow-ups, the clinician can determine the efficacy of the proposed drug or rehabilitation therapy and, above all, the therapeutic compliance of patients to the treatment. The medical team will also be able to contact the patient remotely in order to recommend investigations or instrumental examinations, which the subject will submit at the following face to face meeting.

Thanks to Digital Medicine, patients need to access hospital sites less frequently, reducing their exposure to potential contagion, especially during the COVID-19 pandemic emergency period.

Telerehabilitation for Musculoskeletal Physiotherapy

Telerehabilitation is a real opportunity for patients who need individual motor re-education of a defined anatomical segment - such as a limb - for orthopaedic or musculoskeletal problems. The physiotherapist can guide patients so that they can perform exercises remotely, recover muscle strength or work on the active mobilization of a joint. The rehabilitation of neuromotor disorders, such as balance and gait problems, is instead more difficult. For example, exercises to improve static and dynamic balance should always be offered in the presence of a healthcare professional to reduce the possible exposure to falls for patients. The risk of falling is also increased during postural transfer or walking training exercises. During telereha-bilitation programs, physicians should define the level of computer literacy for patients, especially if they should perform specific motor tasks independently and be aware of potential risks. Patients with long COVID or post-COVID syndrome present a multiplicity of symptoms due to involvement of multi-organ pathology. For these complicated patients, the physician cannot prescribe a standardized reha-bilitation protocol.

The physiatrist should carefully evaluate patients with regards to cardiological, pneumological, neurological, and muscular disorders to set up an individualized rehabilitation protocol.

In most cases, these patients can benefit from treatment including both face-to-face exercises and remote treatments. During the outpatient sessions, the physiotherapist can propose balance exercises and gait training. Instead, during the online sessions, muscle-strengthening or educational activities can be performed.

Telemedicine represents also an opportunity for COVID-19 patients admitted to a hospital far from their home. After the dismission, patients can contact the hospital team, which will provide teleconsultation services to support fellow physicians who are in physical proximity to patients.

Therefore, even in rehabilitation programs (Fig. 9.1), telemedicine represents an integrated approach complementing residential face to face meetings, without com-pletely replacing it.

Fig. 9.1 *Telerehabilitation* station of the "Luigi Sacco" University Hospital in Milan, Italy. The *United States Agency for International Development* (USAID), with the support of *AVSI Foundation*, financed the purchase of a telerehabilitation station and 12 home kits for post-COVID patients discharged from the "Luigi Sacco" University Hospital (Asst Fatebenefratelli Sacco) in Milan, Italy. The physicians and the physiotherapist of the rehabilitation team connect with the patient through the working station. The patient, based at home, uses the home kit received from the hospital once discharged. The home kit includes a tablet, numerous sensors for motion moni-toring, a spirometer and an oximeter for monitoring oxyhaemoglobin saturation. Telerehabilitation allows clinicians to establish ongoing contact with the patient, detecting persistent disorders result-ing from COVID-19 and planning subsequent clinical and instrumental investigations

Does the Patient "Feel Cured"?

What about the patient perceptions? Do patients really feel they are appropriately treated or rehabilitated? Despite the efforts that the physicians and physiotherapists put in during the remote evaluation or treatment process, patients may feel this to be a low-quality rehabilitation service. This subjective perception could have repercussions both on compliance with the proposed treatment and on the essential relationship of trust that should exist between the healthcare professionals and the patient.

For this reason, during remote connections, it is therefore essential to keep a professional attitude regarding the language used and the non-verbal communication. Healthcare professionals should be aware that both verbal and nonverbal communication are essential in order to establish a durable empathic relationship with their patients, especially at a distance.

A satisfaction questionnaire at the end of a telemedicine visit or a telerehabilitation therapeutic cycle can be beneficial feedback from the patients. In this way, it would be possible to evaluate the overall experience, including also the perception of the technology, of the devices and the approach used by the healthcare professionals during the visit.

A detailed checklist with examples is available at the end of this chapter in the "Practice time" section.

Prior to the pandemic emergency caused by COVID-19, Cheshire et al. [6] analysed how stroke patients perceived the use of telemedicine. The researchers handed the recruited patients a questionnaire called *Consultation and Relational Empathy* (CARE) to quantify how patients perceived relational empathy within the therapeutic relationship [6].

There was no statistical difference in the empathy stroke patients felt treated through telemedicine as opposed to those who were offered face to face care [6]. For some patients, therefore, it is possible to successfully activate this new therapeutic and rehabilitative approach, driven by a high quality of service offered, no matter the format.

Possible Solutions and Methodologies

Physicians, nurses, or physiotherapists should have a systematic methodology when using telemedicine or executing telerehabilitation sessions, minimizing technical problems. Healthcare professionals will thus be able to focus exclusively on the relationship with patients, on the language used and on the interaction with the individual on the other side of the webcam.

If a well-defined operational checklist is in place, HCPs can avoid potential challenges such as:

- *IT/technology* (such as connection or data transfer difficulties, or problems with the IT setting or the web platform),

- *logistics* (such as inadequate setting for the visit, for example with poor lighting conditions or insufficient space for the patient),
- *red-tape* (such as the collection of previous clinical documentation or the reports of instrumental investigations, the production of reports or the transmission of these to patients).

Training moments will then have to be structured to strengthen and refine communication and relational skills of the HCPs involved, also by educating professionals to this new approach in conducting health care acts.

The need to implement and enhance telemedicine and telerehabilitation services imposed by the pandemic crisis represents an essential lesson for the future.

Appropriate clinical practices should be clearly defined while favouring empathic communication with patients, even in virtual environments. It is also essential to transmit professionalism and know-how, as well as to clarify the role of Digital Medicine in the diagnostic process and then in the treatment. The crucial point is to establish when Digital Medicine represents a real opportunity for improvement and when, instead, it should be replaced by an in-person evaluation. Digital medicine (complemented with telemedicine and telerehabilitation services) could be a diagnostic and therapeutic opportunity not only during the COVID-19 pandemic crisis, but also in the future.

The Patient Perception: Is the Cure Real or Virtual?

Florence Nightingale, in 1867, concluded by saying: "... *But it is no use to talk about the year 2000*[1]". Only today, in 2020, that model begins to materialize, thanks to the new innovative technological tools allowing healthcare professionals "to get closer" to patients and their homes, thanks to telemedicine, telemonitoring and telerehabilitation.

It is now the right time to discuss and analyse new languages and new methods to conduct patient visits and anamnestic interviews. With an accurate checklist and a *Narrative Medicine* communicative approach offered by a well-trained healthcare professional, there could be no gaps for patients during these remote relationships. Physicians will have to learn and implement good clinical practices, communication strategies, *Narrative Medicine*, and an empathic relationship with the patient in their daily clinical activities. These characteristics together with their high potential will become increasingly important in remote clinical practice. Thanks to this specific training, healthcare professionals will be fully aware of the challenges telemedicine implies and how to tackle them. Only thanks to these specific skills will telemedicine become a real opportunity for treatment. Thus, the patient will be able to experience a "*real*" treatment path, albeit in a "*virtual*" visit or treatment setting.

Practice time

Try to simulate conducting a telemedicine visit, considering all the aspects listed in the following checklist.

CHECKLIST FOR THE TELEMEDICINE VISIT

Before the visit

☐ Check the IT setting at your facility and at the patient's home.

☐ Send informed consent to the online visit, audio-video recording, and the health procedures that must be carried out.

☐ Send patients some demonstration videos to prepare them for the evaluation they will have to undergo.

☐ Ask patients to e-mail all previous clinical documentation and clinical tests performed previously.

☐ Send patients a checklist to prepare for the remote visit.

Preparing the patient for the visit (checklist for the patient)

Room

☐ Quiet and peaceful room.

☐ The room must be sufficiently illuminated.

☐ Sufficiently large room, with no furniture or furnishings that could get in the way.

☐ A bed and chair must be available in the room.

Webcam position

☐ Fixed and stable.

☐ The shot must show the whole patient.

☐ There must be no windows or lights behind the patient.

Patient clothing

☐ The patient should wearcomfortable clothing.

☐ Clothing must be easy to remove to inspect body parts affected by signs or symptoms.

Relatives, caregivers or home assistants

☐ During the visit, a person should assist the patient and move the webcam if necessary.

Access test

☐ The healthcare professional and the patient must test access to the telemedicine portal.

☐ Test the webcam and microphone operation.

Visit

☐ Check the quality of the connection.

□ Ask patients if they can see and hear the directions well.
□ Announces that, when registration starts, informed consent to processing personal data and registration will be requested again.
□ Start recording.
□ Request informed consent from the patient, both for the processing of personal data, for the clinical procedure (e.g. Telemedicine), and the audio-video recording.
□ Complete the anamnestic history collected with the clinical documentation.
□ Proceed with the physical examination.
□ Administer questionnaires or rating scales to estimate the patient's disability.

After the visit

□ Produce the report and send it to the patient (it must be specified that the visit was performed remotely, with which platform and that the patient has given informed consent).
□ Send patients a satisfaction questionnaire.

Try to answer the following questions

1. What limitations would such a visit present?

2. What risks can the patient incur during this type of evaluation?

3. What is missing from this checklist?

4. What language should be used to convey empathy to the patient?

5. What questions could be included in the satisfaction questionnaire sent to the patient at the end of the evaluation?

Call to action

Propose a satisfaction questionnaire to be presented to patients visited with telemedicine.

References

1. Skeet M. Florence Nightingale - A woman of vision and drive. World Health Forum. 1988;9:175–7.
2. Istituto Superiore di Sanità. Indicazioni ad interim per servizi assistenziali di telemedicina durante l'emergenza sanitaria COVID-19. *Rapp ISS COVID-19*. 2020; 12.
3. Tesio L, Buzzoni M. The illness-disease dichotomy and the biological-clinical splitting of medicine. Med Humanit. 2021;47(4):507–12. https://doi.org/10.1136/medhum-2020-011873. Epub 2020 Sep 29.
4. Ministero della Salute. Telemedicina - Linee di indirizzo nazionali. 2012.
5. Epstein NE. Case report (Precis): Two telemedicine consultants miss foot drop: When to see patients in person. Surg Neurol Int. 2020;11:1–2.
6. Cheshire WP, Barrett KM, Eidelman BH, et al. Patient perception of physician empathy in stroke telemedicine. J Telemed Telecare. 2020; https://doi.org/10.1177/1357633X19899237.

Rethinking Sustainability in Healthcare in Times of COVID-19

10

Albina Vegel

> *We do not inherit the earth from our ancestors, we borrow it from our children.*
>
> *Native American saying*

In a recent interview Jane Goodall, a world known primatologist and environmentalist, was encouraged to express her opinion about COVID-19 recent developments and this is what she answered: *"We basically brought this on ourselves by our disrespect of the natural world, forcing animals closer to people, making it easier for a pathogen to jump from animal to a person"* [1]. This pandemic should be a lesson for all of us that we live in a global and connected world. Our human actions affect the nature and have influence on all of us humans, animals, biosphere, our planet and in the end on the health of all of us.

World Health Organization (WHO) declared a COVID-19 global pandemic on 11th of March 2020. Globally, as of 13th June 2021, there have been 175,306,598 confirmed cases of COVID-19, including 3,792,777 deaths reported to WHO. As of 10th June 2021, a total of 2,156,550,767 vaccine doses have been administered. We find ourselves in June 2021 and the world is still slowly but vehemently trying to recover from the health pandemic but not just health also environmental, social, and economic consequences will continue to challenge the world in the upcoming months or even years to come due to the environmental changes that we provoked in this Anthropocene epoch.

A. Vegel (✉)
Universitat Oberta de Catalunya, Barcelona, Spain

The rise of existing and new infectious diseases such as COVID-19 and environmental deterioration, made us think back to the Australian fires in 2020 for example, are a clear evidence that the Planet Earth is in a constant overshot. The health, environmental, social and economic impacts of climate change are visible all around the world. Climate change affects health directly by changing our fundamental needs such as the needs for clean air, safe drinking water, proper sanitation, food supply and our housing. Due to global warming temperatures are rising and because of that many diseases such as COVID-19 and others like yellow fever, malaria, diarrheal diseases, cholera and rotavirus infections are on the rise. It is estimated that between 2030 and 2050 climate change is expected to cause some 250,000 additional deaths a year from malnutrition, malaria, diarrhea and heat stress. The World Bank has reported that climate change could affect more than 100 million additional people living in poverty by 2030 [2].

We need to transform our existing ways of living as a society. If not, there will be no Planet Earth left for the future generations to come. The growing aim of media, governments, international organizations, academics and scientists to talk and write about climate change are not enough anymore. We need a more global, practical, and holistic approach to these matters. The healthcare sector and its workers are one of the crucial actors in making this shift from awareness to practice. Healthcare systems around the world need to rethink their strategies, and especially interms of Agenda 2030, how to achieve Sustainable Development Goal 3 and in what way they can contribute to a more sustainable healthcare and society. In the end all these environmental changes in the world are seen in healthcare institutions where healthcare professionals need to treat humans affected by these environmental changes. But it is not about environmental changes but also economic and social ones that are affecting us. Therefore, sustainability should be a long-term solution in healthcare and all its factors: social, economic, and environmental need to be taken into consideration in order to achieve health equity around the globe. The importance of sustainability is and should be on the rise in healthcare because in post COVID-19 times the need to access health care services could potentially even increase due to forgone healthcare during the pandemic, incrementation of chronic diseases and health post effects provoked after recovering from COVID-19. It may be that during the COVID-19 health crisis sustainability was pushed aside and became a side actor on a healthcare world stage, but it should never stay in the shadows. It should reappear as the main actor on a global stage, given as much of the limelight as possible.

This chapter intents to build an argument about why sustainability is important and should be on the rise in healthcare systems by (1) presenting a Sustainable post COVID-19 healthcare mindset (2) defining what is Sustainability and Sustainable Development, (3) defining Sustainable Healthcare, (4) reviewing United Nations Sustainable Development Goals as a guideline to a better post COVID-19 Healthcare (5) and presenting a practical case of a Sustainability Model by National Health Service in the United Kingdom that can be of use to other healthcare institutions around the globe to re-strategize their healthcare system during or post COVID-19 pandemic in order to confront their future challenges and turn them into more sustainable solutions.

Sustainable Post COVID-19 Healthcare Mindset

In any country around the world the public healthcare system is the one who should provide healthcare for all individuals no matter what their economic and social background is. However, the COVID-19 pandemic has uncovered the failures in the healthcare systems around the world while coping with a new infectious disease on all levels such as disease testing, surveillance, contact tracing and vaccines. All countries around the world, high-income, middle-income, and low-income, were struggling with this new unpredicted situation that they have never ever seen before. Global pandemic COVID-19 affecting so many people around the world was something completely new to all of us. This more globalized world presents us with more globalized challenges and solutions. But no matter if the world becomes more and more connected, in these rapidly changing times in a time of a pandemic each country was fighting alone its own battle against the virus. The global connectedness seemed to be there but, in reality, it appeared to be there more in theory than in practice. Each country kind of stood alone. Let's look at some specific countries as examples.

USA is one of the highest-income countries and has one of the best healthcare systems in the world and it has even received an excellent rating of being a number 1 country out of 195 in the Global Health Security Index (GHSI) but, until today 13th June 2021, it is the country whose COVID-19 numbers rank the highest in total cases and deaths. So far according to Worldometers, they have reported 34,315,923 cases and 614,955 deaths [3]. But even though the country has an impressive public and private health coverage it was not able to successfully manage and retain the spread of the virus.

Let us move further to the European continent and analyse the UK, Spain and Italy. The UK on a Global Health Security Index scored second best place after the USA, but it is also one of the high-income countries whose battle against COVID-19 was unfortunately not a success story. The UK so far has reported 4,558,494 total cases and 127,896 total deaths. In the Worldometers for Corona virus it is in the seventh place for not tackling the disease well. Italy is also one of the high-income countries and was one of the worst affected countries around the world due to COVID-19. The country finds itself in the 31st place on a Global Health Security Index and, so far, it has reported 4,243,482 total cases and 126,976 total deaths which ranks it eighth place in the Worldometers for Corona. Spain is also a high-income country with its healthcare system ranking on a Global Health Security Index at 15th out of 195 countries but also struggled immensely during the COVID-19 pandemic. They had 3,733,600 total cases reported and 80,501 total deaths which ranks the country tenth in Worldometers for Corona ranking.

Let us now move to the Asian continent. India for example on a Global Health Index Security ranking scored 57th place and is one of the lower middle-income countries. On the Worldometers it is reported that so far, they have reported 29,453,298 total cases and 370,626 total deaths which ranks India second right after a highly developed country USA.

What is striking about these numbers is that if we look at the first ten countries that reported the most total cases and deaths on Worldometers the number of cases are from most to least (number 1 being the worst affected): (1) USA, (2) India, (3) Brazil, (4) France, (5) Turkey, (6) Russia, (7) UK, (8) Italy, (9). Argentina and tenth (10) Spain. From this we can see that there is a mix of middle- and high-income countries whose healthcare systems are public, private or a mix of both. In all countries the healthcare system is expected to provide health care for all and with that contributing to the well-being of all its citizens. Helping to protect the health should occur in all spheres of a society without thinking of the financial and social circumstances of individuals. But the COVID-19 pandemic directly exposed failures in the national health systems from when it started. Each country was, or is still, combating this virus in the best possible way and none of these supposedly well developed and high to medium income countries was a good case of practice combating against the virus. All healthcare systems were struggling and trying to manage this new and never before seen situation no matter how well developed and financially strong they were.

This pandemic should be a lesson for all of us that live in a connecting world. How we as humans act towards the nature has an influence on us, animals, biosphere, our planet and in the end, on the health of all of us. One of the lessons to be learnt from this pandemic is that we should invest in health for all citizens of the world on a personal, national and global scale. We could even say that each healthy individual contributes to the healthy planet as an overall. No matter from where you come or where you live. COVID-19 has emphasized the importance of social, environmental, and economic justice for all. These are also the main categories of a sustainable development. We should aim towards a more healthy, inclusive, and equal society around the world. Health and access to health services should be accessible to everyone and represent a universal right. Therefore, the pandemic should serve us now to rethink our current healthcare strategies and the importance of sustainable health systems that could contribute to a better care already now and for the future generations to come. As a native American proverb wisely says:" *We do not inherit the earth from our ancestors, we borrow it from our children.*" With this thought in mind, we should see the healthcare systems as sustainable healthcare systems that ought to contribute to the sustainable development and Sustainable Development Goal 3 and Agenda 2030 which advocates the preservation of our planet for the future generations to come.

For sure the pandemic has exposed failures in the healthcare systems, but this should not slow down the strategies to move towards a more sustainable healthcare but on the contrary it should help us to rethink and reconsider how the healthcare systems around the world could rebuild themselves into a more sustainable friendly healthcare system. It should drive us to move forward into a future that will be based in a more sustainable way in general and in healthcare.

What Is Sustainability and Sustainable Development?

It is very challenging to define Sustainable Development and Sustainability. One of the world's political leaders Kofi Annan in the past recognized that Sustainable Development is a complicated but an important challenge of today's world. These were his thoughts about Sustainable Development concept: *"Our biggest challenge in this new century is to take an idea that sounds abstract—Sustainable Development—and turn it into reality for all the world's people"* [4].

To better define Sustainable Development and Sustainability we need to first look the other way around: at what is *NOT* sustainable. Global Footprint Network [5] invented metrics that measure how much nature we have and how much nature we use. They are warning us that humanity is surpassing its budgetary limit. Their data indicates that carbon emissions combined with all other human demands on the biosphere consume more than 170% of what the Earth replenishes. This means that humanity's ecological footprint corresponds to 1.7 planet Earths. The Earth has been in ecological overshoot ever since they started to measure it in 1969. In the last decades, Humanity has used more resources and generated more waste at a faster pace than the ecosystem could renew. This over-exploitation of the Earth's natural resources clearly demonstrates unsustainable actions towards our planet, suggesting that we should consider ways to minimize these effects as much as possible.

The most used definition of Sustainability today is from the 1987 Brundtland report written by United Nations World Commission on Environment and Development titled Our Common Future, defining Sustainable Development as: *"Development that meets the needs of the present without compromising the ability of future generations to meet their own needs"* [6].

Richard M. Clugston in the foreword of Higher Education and the Challenge of Sustainability described Sustainability in a similar way as: *"a new approach to social and economic development and global security which integrates concerns for short term economic gain with concerns for future generations, cultural and biological diversity, and social well-being"* [7].

Sustainability is often thought of as a long-term goal to a more sustainable world, while Sustainable Development refers to the many processes and pathways to achieve this long-term goal. Throughout the last decades many definitions have been created and this has led to a confusion in understanding and a lack of consensus about the concept among scholars [8–11].

Sustainable Development is normally represented by *three main integrated dimensions: environmental, economic, and social.* Giddings, Hopwood & O'Brien [12] presented these perspectives in a simple three-ring model. This way of describing Sustainable Development by the three equal dimensions of the environmental, economic, and social dimensions is often referred to as the three pillars of Sustainable Development. This model has been criticized to give less importance to environmental dimensions in relation to economic and social dimensions [13]. Also, other ways of understanding the relationships between the dimensions have been suggested [12, 14], yet they are less prominent because during the last decades after the Earth Summit in Rio this three-ring model has been widely accepted in international environments and international organizations such as the United Nations. Most contemporary definitions of Sustainability share at least three core elements, despite their diverse emphasis and perspective:

– *First, most definitions are in line with the three pillars of Sustainability: the economic, social, and environmental.*

- *Second, many definitions of Sustainability emphasize that innovation should extend beyond the existing standards.* Sustainability efforts are depicted as exceeding the basic environmental, social, and economic laws and regulations.
- *Third, Sustainability often focuses on intergenerational equity and fairness.* This is formed in an idea that our present actions must account for what impact they will have on others and especially for the generations that will follow [15].

Despite its broad use in many areas up to this current day, Sustainable Development and Sustainability are still difficult to define. There are still many challenges to overcome as these concepts represent different things to different people and further explorations of definitions are needed. But in a fast-paced changing world, this should not be seen as a disadvantage but on the contrary as an advantage. New developments and research can occur daily and can contribute to further understanding and development of the concepts. But one thing is for sure: the existing definitions offer a starting point for a conversation among many disciplines to join their forces and guide them to a common vision of the future. Healthcare systems and institutions are organizations that can substantially influence further development of the concept of Sustainable Development and contribute to the long-term Sustainability around the world.

Sustainable Healthcare

While all these crucial talks of how to be more sustainable were running around the globe a COVID-19 pandemic hit. When the pandemic hit there was no time to think it all through but what was the most important was to save lives and react as quickly as possible. One of the consequences of these rapid reactions to COVID-19 are also extra medical waste that has been produced in the last year and a half: from face masks, personal protection equipment to gloves and more recently the needles from the number of vaccine doses that are distributed in the population. What to do with all this extra waste?

A sustainable approach to health systems is necessary now and in a post-COVID-19 times because this kind of approach can provide us with health care services that will remain safe and assure us the highest quality. This approach will also look after a whole population and will be more inclusive for all. Especially for the most disadvantaged who were the most vulnerable during the pandemic and can remain in this position even after the pandemic is over. Health systems should not take care just for "repairing" the body but nurture and care for it in a holistic way, meaning that health systems should recognize other processes and factors that influence our health at the same time such as economic, environmental and social factors that are tightly connected with sustainability aims.

To be able to further understand and implement quality sustainable changes in a healthcare we need to define what is sustainable healthcare. Rio Declaration 1992 [16] about environment and development in Principle 1 speaks of sustainable healthcare as: *"human beings at the central concern of sustainable development*

(...) living a healthy and productive life in harmony with nature. Centre for Sustainable Healthcare from United Kingdom describes sustainable care as *"care that is delivered in a way that does not adversely affect the health of the population and does not use resources in a way that prevents tomorrow's health professionals from providing care"* [17]. The World Health Organization speaks of sustainable healthcare as *"a system that improves, maintains or restores health, while minimizing negative impacts on the environment and leveraging opportunities to restore and improve it, to the benefit of the health and well-being of current and future generations"* [18]. Amarantha Fennel-Wells [19] in her presentation about what is Sustainable Healthcare categorizes it in these five features:

- Considers the needs of the entire population.
- Supports illness prevention.
- Builds health and resilience.
- Highlights resource use and management.
- Takes financial, environmental, and social aspects into account.

We could go even further and say that a healthy human being represents a healthy planet and the other way around. Climate change is damaging human health today and will have an even greater impact in the future. In 2015, the Rockefeller Foundation and the Lancet launched the concept of Planetary Health as the Rockefeller Foundation-Lancet Commission. Planetary Health is a concept that introduces this idea meaning that *"the health of human civilization and the natural systems on which it depends leads us to planetary health"* [20] and should be aimed to be nurtured and preserved.

What is clear is that sustainability is an important concept for today's world that encounters itself in a COVID-19 pandemic. But not just for today but of a great importance for the future of us all.

It can be adapted to many working fields and sectors. But it should not just stay written in theory books. It should be adapted to as many practical projects and strategies around industries and especially in healthcare, the industry which can contribute to a great improvement in health of humans, their wellbeing and even promote planetary health.

A study Health Care's Climate Footprint from 2019 [21] is one of the first studies that investigates healthcare climate footprint and outlines a set of actions the sector should follow to act in accordance with a more sustainable healthcare goals. But it also presents us with striking global statistics for EU and 43 countries around the world. The highest contributions to the global health care climate footprint came from the United States (546 million metric tons of CO_2e), China (342 $MtCO_2e$), and the European Union (248 $MtCO_2e$). But if healthcare were a country, it would be the fifth largest contributor to the contamination of the planet. Its global climate footprint can also be compared to the greenhouse gas emissions from 514 coal-fired power plants.

Healthcare's given mission to protect and promote health is also in the core interest of sustainability. Healthcare needs to recognize the importance and responsibility of adapting to sustainability. Even the Hippocratic Oath which is the core commitment for medical professionals says "first, do no harm". This tendency should also be adapted to climate footprint and other sustainability initiatives in the healthcare sector and with that influencing other sectors to contribute to this change and try to reach "net zero" around the year 2050. But it is not just about the industry it is also about the responsibility of its people [21].

Between December 2019 and January 2020 in Northern Virginia in United States 16 hospital staff were asked about the connection between sustainability and health. Qualitative study intended to evaluate how health personnel react to the proposal that sustainable solutions are also health solutions. The sustainability essay was given to them to read with six categories of solutions for hospital implementation like energy, food, transport, building, consumption, and actions that can be taken in surrounding cities and communities. One of the questions they were asked was: Do you see a connection between sustainability and health? Ten participants emphasized pollution, climate change, and other climate impacts as threats to public health and six participants emphasized hospitals' production of waste, greenhouse gas, and other environmental impacts. When asked do you see environmental sustainability as a part of your job, eight participants gave a definite positive answer, three gave a definite negative answer, and five said that they have a moral responsibility to be sustainable, but that responsibility is separate from their profession [22].

From this research we can see that health professionals have a tremendous opportunity to become important advocates for sustainability solutions that could improve public health. From a recent YouGov study from 2021 it shows results that in the UK medical doctors and nurses are in the top five most respected professions: nurses came fifth, and medical doctors in second place [23]. Positioning sustainability solutions as health solutions should be a viable method amongst health professionals and see them as well as sustainability defenders in their place of employment. But health professionals also have a potential to inspire sustainability efforts externally with the public and policy makers, as well as improving sustainability within the health-care industry, which is responsible for a substantial fraction of the world's air pollution and greenhouse gas emissions. Healthcare workers can be true ambassadors of sustainability.

Each nation's healthcare system directly and indirectly releases greenhouse gases while delivering care and procuring products, services, and technologies from a carbon intensive supply chain. In recent years health care climate footprint estimations were explored in some countries. Two studies in the United States found the country's health care emissions to alternately have reached 8% and 9.8% of the national total, respectively, with the latter estimate comprising 655 million metric tons of carbon dioxide equivalent. In the United Kingdom, the National Health Service (NHS) and Public Health England estimated the health and social care climate footprint in England in 2017 to be 27.1 metric tons of carbon dioxide, representing around 6.3% of the country's climate footprint. Similar studies had comparable findings in Australia 7% and Canada 5%.

These measurements are promising in order to advance with the minimizing of carbon footprint, but it is only a start. Still much more should be done in the future to adapt to sustainable changes and lower carbon footprint in post COVID-19 times in healthcare [21].

Sustainable Development Goals for Post COVID-19 Healthcare

"Health is a precondition for and an outcome and indicator of all dimensions of sustainable development." (United Nations 2012. A/RES/66/288. The Future We Want).

The 2030 Agenda for Sustainable Development adopted by all United Nations Member States in 2015 provides a shared blueprint for peace and prosperity for people and the planet, now and into the future. At the core of the Agenda 2030 are 17 Sustainable Development Goals, which are an urgent call for action by all countries in a global collaborative partnership. These goals aim at ending poverty and other deprivations that go closely hand in hand with strategies to improve public health and education, reduce inequality and encourage economic growth. All these goals should respect climate change and work towards preservation of nature.

In this strategy one of the core fields is also health that is covered in Sustainable Development Goal 3, and it is formed as "Ensure healthy lives and promote well-being for all at all ages". This aim is important because ensuring healthy lives and promoting well-being is crucial to building prosperous societies. It is useful to see the relationship between health and sustainable development in three ways: health as a contributor to the achievement of sustainability goals, health as a potential beneficiary of sustainable development and health as a way of measuring progress across all three pillars of sustainable development policy [24].

There are many emerging challenges to overcome for 17 Sustainable Development Goals to name just a few of Sustainable Development Goal 3:

- *Major shifts in the age structures of countries.* Low-income countries have unprecedented numbers of people aged under 24, whereas in a high- and middle-income countries we are faced with an aging population.
- *All countries need to develop capacities and to think creatively and innovatively to deliver health and wellness.* Not just in each of the life stages but also for every person throughout life, regardless of the socio-economic, health, gender, and any other defining status.
- *New infectious diseases appear frequently.* For example, the infamous COVID-19 pandemic that spread globally and requires continuing development, maintenance and prioritization of national and global public health institutions, data collection, analysis, and technical capacities [25].

To sum up, there will always be sustainable opportunities and challenges to overcome and as Lao Tzu wisely said: *"the journey of a thousand miles begins with a single step"*. 17 Sustainable Development Goals are not just a single step but a roadmap of many pathways, ideas, and strategies to immerse ourselves into the adventure of Sustainable Development and Sustainability and start changing the world today for us and the future generations to come. It is also good for the Healthcare systems of the world to use as a guideline for a transformation towards a more sustainable healthcare model.

Another good practical case in this sustainable trajectory is also United Kingdom National Health Service that invented a Sustainability Model to help with managing sustainable innovations in a healthcare system that is explained in the following section.

Post COVID-19 Sustainable Ideas and NHS Sustainability Model

United Kingdom's National Health Service (NHS) reduced the health and social care climate footprint by 18.5% since 2007. Its goal is to comply with the country's Climate Change Act, which sets a requirement of reducing the footprint further so that United Kingdom achieves a 34% reduction by 2020 and an 80% reduction by 2050. There are other outstanding local and regional examples in Europe, particularly in Scandinavia and the Netherlands, where zero emissions hospital buildings, increasing organizational commitments to carbon neutrality, innovative climate-smart technologies, and strategies to address supply chain emissions are being adopted in the sector. But for the reference of best practice for the need of this chapter British public health sector NHS will be used.

Report from April 2021 Creating a Culture of Sustainability: Leadership, Coordination and Performance Measurement Decisions in Healthcare written by Tonya Boone defined key organizational elements that support the effectiveness of the sustainability initiatives in healthcare:

1. *There must be an executive champion who advocates for sustainability to important stakeholders and a sustainability coordinator who oversees day to day activities.* Sustainability coordinators have a variety of backgrounds. Critical coordinator capabilities include familiarity with environmental management systems, communication skills, process analysis and management skills and a strong commitment to ongoing learning. They utilize a variety of means, including best practices and processes, information technology networks, and green teams to assure flexible systems, communication, and commonality of approach at all levels [26].

   NHS has an experienced Chief Sustainability Officer, Dr. Nick Watts, who stepped down after 8 years of being executive director at the Lancet Countdown to join the NHS in October 2020. The Lancet Countdown is the world's leading publication about health and climate change with 2020 published reports, 43 indicators, 120 collaborating experts and 35 collaborating organizations [27]. At the NHS Dr. Nick Watts is responsible for the NHS's commitment to delivering a world-class net zero emission health service and to leading the nationwide Greener NHS team.

2. *Values of the sustainability initiative must be reflected in the organization's high-level mission, vision, or values.* This linkage helps us to connect sustainable operations with patient care and define clear organizational norms and decision-making [26].

The NHS has a clear mission to provide health and high-quality care for all. Its vision is "to deliver the world's first net zero health service and respond to climate change, improving health now and for future generations." [28] It has a clear strategy connected with climate change and sustainability. In October 2020, the NHS adopted a multilayer plan with deliverables and targets to become the world's first carbon net zero national health system. NHS specific targets to achieve net zero emissions are:

- For the NHS Carbon Footprint (emissions under NHS direct control), net zero by 2040, with an ambition for an interim 80% reduction by 2028–2032.
- For the NHS Carbon Footprint Plus, (which includes our wider supply chain), net zero by 2045, with an ambition for an interim 80% reduction by 2036–2039.
- NHS defined deliverables are:
- New ways of delivering care at or closer to home, meaning fewer patient journeys to hospitals.
- Greening the NHS fleet, including working towards road-testing a zero-emissions emergency ambulance by 2022.
- Reducing waste of consumable products and switching to low-carbon alternatives where possible.
- Making sure new hospitals and buildings are built to be net-zero emissions.
- Building energy conservation into staff training and education programs.

3. *Performance measurement systems are used to monitor and guide sustainability activities.* The organization has carefully defined its sustainability performance indicators and is regularly monitoring their progress at the highest organizational level. These same indicators are used on a global as departmental level of the organization [26].

One of these diagnostic measurement tools at the NHS that can be used on a global or departmental level is the Sustainability Model that will identify the strengths and weaknesses in an implementation plan and predict the likelihood of sustainability for an improvement initiative.

In the following section the functionality of the NHS Sustainability model is presented [29].

Sustainability is not just an add-on in a business strategy but needs to be a clear result of an effective preparation and implementation. To hold a productive discourse about it would be good but only half as good as solutions with a solid communication, plan and strategy. Initiatives need to be written down, planned, reassessed and result in concrete action plans.

NHS Institute for Innovation and Improvement (Lynne Maher) in collaboration with University of Wisconsin (David Gustafson and Alyson Evans) elaborated a Sustainability Model and Guide for their internal needs but it could be also adapted to any other healthcare organization to improve their sustainability strategies. The *Sustainability Model is a diagnostic tool that will identify strengths and weaknesses in any implementation plan in healthcare and predict the likelihood of sustainability for an improvement initiative.* Sustainability Guide is a practical guide with advice on how it is possible to increase the likelihood of sustainability for an improvement initiative and it was developed as a direct result of requests from NHS staff who were using the Model.

The NHS Sustainability Model is an easy-to-use tool which aims to help NHS improvement teams to:

- Self-assess with ten factors criteria for sustaining change.
- Recognize and understand key barriers for sustainability, relating to their specific local context.
- Plan for sustainability of improvement efforts.
- Identify strengths in sustaining improvement.
- Monitor progress over time.

The Model does not provide a holistic strategical plan on how to start implementing sustainability from scratch. Therefore, you cannot use it to assess whether a department, organization or a health community is to sustain a change in general. It is more practical, and its application needs to be more specific. It is designed to be used in an already planned or ongoing improvement initiative or project such as using it within a health community to work on an improved approach to how to provide more quality care for the elderly in a retirement home. The Sustainability model can be used by individuals or teams and in a departmental or a global level. But it is advisable that more members of the team are included in this assessment since with the collaboration of more people you get a more general and objective perspective of the needs to be improved in a project. The model is advised to be used in different phases of implementations to give an idea of what has already been done but also to think of what can still be further improved. The model can be used at any desired time that is convenient to improve the initiative, but these timelines were suggested as an example of good practice: at the first planning stage, at initial pilot testing and a few weeks after the improvement has been implemented.

The model itself will guide you through 3 main categories and 10 factor descriptions which can be elaborated in a questionnaire and then analysed through the score system bar chart or portal diagram. After the result analysis NHS recommends focusing on the first three factor descriptions that ranked the highest to be improved. For a more detailed reference, please, go to NHS.

Sustainability Model and Guide. These are the main categories with its factor descriptions in a *Sustainability Model*:

- *Process:* benefits beyond helping patients, credibility of the evidence, adaptability of improved process, effectiveness of the system to monitor progress.

- *Staff:* staff involvement and training to sustain the process, staff behaviours toward sustaining the change, senior leadership engagement, clinical leadership engagement.
- *Organization:* fit with the organization's strategic aims and culture, infrastructure for sustainability.

It is very promising to see that the public healthcare system as big and influential as the NHS is changing and looking into ways through in which they can adapt to being more sustainable. With these kinds of awareness initiatives, actions, and strategic planning many in the industry can follow. The NHS is a good study case example for them to look to. Of course, there is still a lot that needs to be done in a healthcare sector to be able to improve sustainability initiatives, plan and measure them but this is a good solid start. With continuous innovation and development in healthcare, we can hopefully achieve the reduction of the human impact and preserve our planet for the future generations to come.

Conclusion

Let us conclude with an ancient wisdom of the native Americans that is still very much alive today in the era of COVID-19 pandemic, whose mantra repeats itself through the chapter and it is aligned with sustainable development and sustainability and can even give us a strong wake-up call to reframe our thoughts and actions towards a more sustainable healthcare development. It goes like this: "We do not inherit the earth from our ancestors, we borrow it from our children." Let us keep this thought in our minds while improving the healthcare systems of the world. The aim of this chapter is to contribute and encourage further discourse and action development among healthcare systems and institutions to find new plans, strategies and actions that will be more aligned with sustainable practices. It is important to talk about sustainable development and sustainability, define its concepts and understand the theory behind it but it is even more important to make a shift from its awareness to its action plan. From this chapter it is visible that healthcare institutions are dedicating more and more time and resources to find a more sustainable solutions in its sector even though COVID-19 has disturbed these initiatives. But they are not forgotten. Healthcare sector is beginning to understand that sustainable healthcare solutions are very important and are here to stay, not just for today but for the future generations to come. The healthcare sector should be aligned with sustainability. In today's world even more than ever sustainability is of crucial importance and we must reconnect with nature and respect it to preserve our health, biodiversity, and, with that contribute to a healthier planet for us and for the future generations to come. We need to remember that we only have one planet Earth as a home and that there is no Planet B. If we do not respect nature and treat it with more respect our health, environmental, economic and social challenges might become even more complex than in the last COVID-19 years of 2020 and 2021. We should learn from our previous experiences. All individuals and sectors should pitch in and help but the healthcare sector is directly connected and influenced by these ongoing changes and will hopefully further develop itself towards a more sustainable healthcare.

What Does Sustainability Mean to Me?

Sustainability is the capacity to endure in a relatively ongoing way across various domains of life. In the twenty-first century, it refers generally to the capacity for Earth's biosphere and human civilization to co-exist. But this way of living is not new at all, maybe just the word Sustainability. Around the world this way of living has existed in many different nations, countries, forms, and ways for generations. We should listen to our ancestors and learn from their stories and ways of living. It is not for nothing that the British Statesman, Winston Churchill, speaking in the House of Commons in London in 1948, and, importantly, slightly altering Georges Santayana's words, [30] "Those who fail to learn from history are condemned to repeat it." Lessons from ancestors throughout history may not always be successfully applied but they can provide good insight into the present and the future. Let us listen to the stories of history and reflect upon them.

Ever since I was a child and to this day sustainability has always been a way of living for me. I was just not aware of it or of how to name this way of living. My parents and grandparents taught me how-to live-in harmony with people and nature. Sustainability for me represents a bridge between the past, present and the future. It means peace, tranquillity, relaxation, respecting each other and the nature that surrounds us, paying attention to traditions, and learning from stories and passing on all these values to the people that surround me.

For me it is also a bridge because it connects me to my origins back home in Maribor, Slovenia where I was born and raised. This means that wherever I am I can still show people around the world my values through these sustainable values in which I was raised as a child and developed into a young adult. No matter where I go, I always carry those values with me. I was one of those lucky kids who was able to enjoy living close to the city but at the same time enjoy the multiple perks of living next to the mountain Pohorje and admiring its beautiful nature and changing colours through all seasons of the year. But there is more! Not only was nature at a walking distance from our home, but my family also had and still owns a huge vegetable garden next to our house where I was able to learn how to live with nature and in the different seasons encounter lettuce, carrots, beans, potatoes, tomatoes, peppers, parsley, peas, strawberries and even gaze at beautiful flowers such as tulips or roses.

The "Maribor garden" always accompanies me in some form (Fig. 10.1), and I always try to recreate it no matter where I am (Fig. 10.2). Now, my little

"Maribor garden" can be found in the Netherlands where I can also enjoy homegrown tomatoes, peppers, strawberries, parsley, roses and tulips. Through my garden I can connect different worlds, values and times creating a bridge between the past, present and the future. It is a small act, this garden of mine, towards sustainability but with its example I would like to convey a message:

Each small step matters in sustainable actions, and no matter how small it is, it counts! To conclude, I would like quote from Howard Zinn: "We do not have to engage in grand, heroic actions to participate in change. Small acts, when multiplied by millions of people, can transform the world." [31].

Fig. 10.1 Tulips Garden in Maribor, Slovenia

Fig. 10.2 Photo by Albina Vegel "View from the walk to Pohorje in Maribor, Slovenia"

Practice Time
1. What is sustainability for you?
2. Do you practice sustainability in your day to day life?
3. What are the major challenges to practice sustainability?
4. Mention three things to promote sustainability.
5. What ways could you think in improving sustainability in the health-care system?
6. How to teach sustainability to children?
7. Have we forgotten how to "Gather the flowers but spare the buds" [Andrew Marvell]?

References

1. France24. World needs "new mindset for our survival", says Goodal. 2021. Available from: https://www.france24.com/en/live-news/20210520-world-needs-new-mindset-for-our-survival-says-goodall
2. Connecting the Dots. Strategy Note. HIV, Health and Development 2016-2021. United Nations Development Programme. 2019. Available from: https://www.undp.org/content/dam/undp/library/HIV-AIDS/UNDP%20HIV%20Health%20and%20Development%20Strategy%202016-2021.pdf.
3. Worldometer. Coronavirus cases. 2021. Available from: https://www.worldometers.info/coronavirus/.\
4. Annan K. Foundation. Thought about. Sustain Dev. 2018; Available from: https://twitter.com/kofiannanfdn/status/988808037375856640?lang=en
5. Global Footprint Network. Our past and our future. Available from: https://www.footprintnet-work.org/about-us/our-history/.
6. Our Common Future. Bruntland Comission. Oxford: Oxford University Press; 1987. Available from: https://sustainabledevelopment.un.org/content/documents/5987our-common-future.pdf.
7. Clugston RM. Higher education and the challenge of sustainability. Amsterdam: Springer; 2004. Available from: https://www.springer.com/gp/book/9781402020261.
8. Marshall JD, Toffel MW. Framing the elusive concept of sustainability: a sustainability hierarchy. Environ Sci Technol. 2005;39(3):673–82.
9. Becker CU. The meaning of sustainability. In: Sustainability ethics and sustainability research. Cham: Springer; 2012. p. 9–15.
10. White MA. Sustainability: I know it when I see it. Ecol Econ. 2013;86:213–7. Available from: www.sciencedirect.com/science/journal/09218009/109World
11. Filho WL. Dealing with misconceptions on the concept of sustainability. Int J Sustain High Educ. 2000;1(1):9–19.
12. Giddings B, Hopwood B, O'Brien G. Environment, economy and society: fitting them together into sustainable development. Sustain Dev. 2002;10(4):187–96.
13. Kopnina H. Revisiting education for sustainable development (ESD): examining anthropocentric BiasThrough theTransition of environmental education to ESD. Sustain Dev. 2014;22(2):73–83.
14. Walshe N. Exploring and developing student understandings of sustainable development. Curricul J. 2013;24(2):224–49.

15. Weisser RC. Defining sustainability in higher education: a rhetorical analysis. Int J Sustain High Educ. 2017; Available from: https://www.emerald.com/insight/content/doi/10.1108/IJSHE-12-2015-0215/full/html.
16. Rio Declaration 1992. Accessible available from: https://thefactfactor.com/facts/law/civil_law/environmental_laws/rio-declaration/874/#:~:text=Peace%2C%20development%20and%20environmental%20protection,Charter%20of%20the%20United%20Nations (05/05/2021).
17. Center for Sustainable Healthcare United Kingdom. Available from: https://sustainablehealthcare.org.uk/courses/introduction-sustainable-healthcare
18. World Health Organization about sustainable healthcare. Available from: https://www.euro.who.int/__data/assets/pdf_file/0004/341239/ESHS_Revised_WHO_web.pdf.
19. Amarantha Fennel-Wells. 2020. Sustainable Healthcare: An Introduction with a perspective from Wales. Available from: https://www.youtube.com/watch?v=ee6Gt8wtpQc.
20. Safeguarding human health in the Anthropocene epoch: Report of The Rockefeller Foundation–Lancet Commission on planetary health. Available from: https://www.thelancet.com/commissions/planetary-health.
21. Health Care's Climate Footprint. How the health sector contributes to the global climate crisis and opportunities for action. Arup. 2019; Available from: https://www.arup.com/-/media/arup/files/publications/h/health-cares-climate-footprint.pdf.
22. Hubbert B, Ahmed M, Kotcher J, Maibach E, Sarfaty M. Recruiting health professionals as sustainability advocates. Lancet Planetary Health. 2020;4(10) Available from: https://www.researchgate.net/publication/346119509_Recruiting_health_professionals_as_sustainability_advocates.
23. YouGov. Scientists and doctors are the most respected professions worldwide. 2021. Available from: https://yougov.co.uk/topics/international/chapters-reports/2021/01/28/scientists-and-doctors-are-most-respected-professi.
24. 17 Sustainable Development Goals. United Nations. Available from: https://sdgs.un.org/goals.
25. TST issues brief: Health and Sustainable Development. Available from: https://sustainabledevelopment.un.org/content/documents/18300406tstissueshealth.pdf#:~:text=Health%20is%20central%20to%20the%20three%20dimensions%20of,rights-based%2C%20inclusive%2C%20and%20equitable%20development%20seeks%20to%20achieve2.
26. Healthcare without harm. Available from: https://noharm.org//.
27. The Lancet. Available from: https://www.thelancet.com/.
28. National Health Service United Kingdom. NHS UK. National ambition. Available from: https://www.england.nhs.uk/greenernhs/national-ambition/.
29. Sustainability Model and Guide. Institute for Innovation and Improvement. NHS. Available from: https://www.england.nhs.uk/improvement-hub/wp-content/uploads/sites/44/2017/11/NHS-Sustainability-Model-2010.pdf.
30. Santayana G. The life of reason. New York: Open Road Media; 2017.
31. Zinn H. A power governments cannot suppress. San Francisco: City Lights Books; 2006.

Maybe In, Maybe Out, May Be with the Pandemic

11

Maria Giulia Marini and Jonathan McFarland

> *"…because maybe,*
> *you're gonna be the one who saves me,*
> *and after all, you're my wonderwall"*
>
> *(Oasis)*

While writing these lines (April 2021), some countries such as UK, Israel, USA, the wholeness of Eastern Countries (apart Japan and India) are recovering from the pandemic, while Continental Europe is still having to face a horrifying number of dead people every day. When will we all be vaccinated? Will the vaccine solve the problem of this Covid-19 forever or will we all need to be vaccinated regularly every year? Will we need to fall back into lock down (currently, all Italy is in Red Zone, with closed schools, closed cinemas, theatres, closed restaurants and bars, forbidden to leave our county of residency? New variants of the virus are continuously being discovered, this time is Japanese, named Eek, and it seems to respond less to vaccinations [1]. But virologists say there will be new ones.

Again, fear is spreading in the social media around the world: however, we have to distinguish from what is a useful anxiety, that one restricted to scientists, doctors, nurses and first aid people, trained to face and control their emotions, and what is a useless panic attack, very common of people who unfortunately have not being

M. G. Marini (✉)
Fondazione ISTUD, Milano, Italy
e-mail: MMarini@istud.it

J. McFarland
Faculty of Medicine, Autonomous University of Madrid, Madrid, Spain

Sechenov University, Moscow, Russia

taught to live in peace with their psyche; people who have difficulties in bunding safety and pleasure space even in a dangerous world [2]. We will later return with some elements to face difficult emotions. Now, let's remain with the beauty of the words as our raft to cross this Odyssey in the open sea. May be. Maybe it can be, May be it cannot be: this word, maybe, belongs to the 64 universal words that compose the Natural Semantic Metalanguage (NSM), and means uncertainty, confusion and is open to hope [3]. This term is spread throughout world cultures as a living proof that certainties in the facts and events of all human beings can never be taken for granted, but at best planned or forecast. NSM is the language which possibly guided the evolution of our ancestors, when facing difficulties, challenges, open endings, the "unknown" amongst possible options, and "maybe "was there then and it is still here now. Let's open the window to the importance of this Natural Metalanguage since these 64 empirically studied words still remain universal. The NSM-list is a summary of words and grammar rules that are common in languages and therefore constitute an external or cultural manifestation of a natural language or semantic ability. Their discoverer, Anna Wierzbicka, Professor at the Australian National University assumes that meaning is mainly given by the grammatical structures in which words are placed; a common human concept is not just an isolated semantic element, but rather, it constitutes a small lexicon-grammatical system.

In this way, this NSM implies the language of thought, and should then logically interconnect with the functioning of the nervous system and brain. So far, the debate is still open as to whether this is not only an evolutionary language but also an inherited one, present in our genetic material, and carrying the genetic code. However, the universal words and the universal grammar of the NSM are there to show our brain's cognitive and emotional patterns.

And returning to maybe, this is not to be meant as a cheap matter of "casting the die", since, if we had been more accurate in listening to the expert voices of virologists in recent years, we would not have arrived where we are today, so unprepared. On the contrary, this uncertainty fostered the extraordinary efforts played by managers in health care, doctors, nurses, and all the volunteers to reformulate the process of care, and boosted the wonderful discoveries of the scientists with these new vaccines, although we do not know how long they will protect us and, speaking openly, in the realms of probabilities, our older generations. Data on mortality ages from Covid-19 are quite clear; up to now, in Italy, one of the oldest countries in the world, it ranges from 81.6 years to 80.3 years [4].

In NSM *maybe* belongs to the "logical concept", together with a couple of very interesting universal term: "*if*" and "*because*".

"If" is a term that has to do with planning and consequences: "*if we had prepared earlier with swabs, earlier lock downs, there was more testing and less infections*"; "*if we were faster with the industrial plant in Europe for producing vaccines, we were not so dependent on UK and USA*". Well, "if" is the "alter ego" of the word "maybe".

Many questions still linger, and they cannot be solved only by science in a neat causal – effect relationship but they open the door to the word "maybe"; some examples: will we ever return to a normal life? What about the forecast *new*

Fig. 11.1 The Unbalance
between Maybe, If,
Because

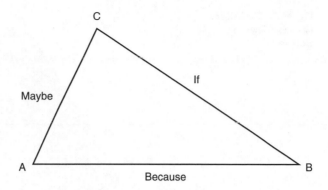

normality? Will we be paid back for the loss of freedom and of for those close ones we have lost? Was this experience a gift, through which we were allowed to discover more of our inner resources? A balance between these logical concepts: "*maybe*", which opens to infinite degrees of freedom in statistical terms, and, is the original meaning of the word chaos (in its etymology it means 'the place and time when every possibility exists'), "if", which limits the number of degrees of freedom, conditions it, and brings some order to chaos, and last but never least, "because", which is the explanation of the cause- effect relationship: "*the pandemic spread because we have almost destroyed biodiversity*", "*the vaccines works because they mimic the viral protein of the spike*." Three dimensions, this triangle in which we can diagnose ourselves to see which sides of the triangles are longer, and, if we wish, move from a scalene (Fig. 11.1), asymmetrical situation to an equilateral symmetric triangle (Fig. 11.2).

It is evident that in the first case we wish to reach immediate fast solutions, search for and accept in haste, some "becauses" that could give some kind of evidence and- maybe- some others full of falsehood. The "maybe" is lived on the side of discomfort, and so we avoid staying in that place, the "if" as a logical pattern is in the middle.

In Fig. 11.2, the beauty of simplicity is given by symmetry, a pattern that we can find in nature. A study of geometric regularity was performed as a simplicity metric for understanding the aesthetics of basic polygons. Undergraduates rated the perceived beauty of triangles varying in geometric regularity defined by their side-length standard deviation. Each type of triangle was used: equilateral, isosceles, right, and scalene. Result showed that ratings of aesthetics increased with an increase in regularity and were highest for symmetric shapes [5].

Flexibility is moving from one triangle to another, from one side to another, from taking part in this cosmic dance, in which we are not the only performers, but accepting that there is a portion of "perhaps", an "open ending" and considering this open end as a possible "*wonderwall*", as in the song by Oasis,[1] an expression in which everyone is free to assign a possible meaning after the reading of this part.

---

[1] Oasis were an English rock band formed in Manchester in 1991; one of the main representatives of the Brit Pop movement.

Fig. 11.2 The balance
between Maybe, If,
Because

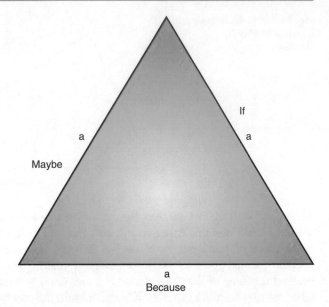

One message I would like to bring is the interdependence amongst people; that nobody saves herself/himself alone, "*maybe you've gotta be the one who saves me*", and adding another jingle of reciprocity "*Maybe I've gotta be to be the one who saves you*" becoming reciprocal wonder walls. Love creates wonder, always inventing something new, able to build a wonderland, but at the same time as strong as a wall, protecting people. *Maybe*. How do you feel now about this word?

The Gift of the Bio-Psycho- Social and Existential Model

The biomedical model can provide cures and vaccines, for which evidence shows that they are starting to protect our older generations, health care providers in the front line of assistance, and us. However, we are not only bodies, but as Aristotle said in his Politics, the man (and the woman I add) is a "*political animal*" because he/ she is a social creature with the power of speech and moral reasoning, and with morality where ethics is encompassed, not merely prejudices of bigotry.

A model with a broader horizon as the biological psychological social and existential one, can include by juxtaposition, the standardized biomedical one, bypassing its reductionism. This model is framed into four domains, that, as in an osmotic process, are constantly "leaking" into the other compartments: the well-being of the body reverberates on the mind, on the social life, the psychological and the core life, that purely existential part made up of ethics, values, and inner beliefs, awe and wonders. In synthesis, any part impacts onto the others, without questioning which part was the first to start [6].

During this Covid-19 pandemic, the WHO issued guidelines to cope with stress, and there is a constant call by the majority of local governments to their citizens to be ans stand resilient. There is so much asking, talking and teaching about *Resiliency,* that somehow, ironically, we are developing antibodies to this word more than to the virus. This is a fact since easy solutions such as this imperative to "be resilient", may work in the very short term with some individuals but probably not in the long run. Our inner Rebel Child is always vigilant when too many commandements are given by authorities.

Let's spend some time to explore to whom we owe the concept of resiliency so that maybe by understanding its roots we can transform this word into more than an external input but as a living action. We are grateful to the ancient philosophers, and in particular to the Stoics, whose founder was Zeno of Citium in Athens in the early third century BC. Stoicism is a philosophy of personal ethics, full of tension towards accepting the reality as it is, and not allowing oneself to be controlled by the desire for pleasure or by the fear of pain. The human mind should be used to understanding the world and to do one's part in nature's plan, and by working together and treating others fairly and justly.

Stoicism was boosted in Rome with Seneca and reaching the Roman Emperor, *Marcus Aurelius,* who, by the way, was also the Commander-in-Chief of the Roman Army, so, certainly not a person detached from pragmatic decision-making. This is probably the reason why stoicism was well appreciated by the Romans and their materialistic culture, which left little room for fantasies and lingering souls in the sky.

Philosopher *Lucius Anneus Seneca* was born in Spain, suffered from severe asthma and experienced tuberculosis when he was young but thanks to his mother's connections, he was able to move to Rome and become Chief Advisor to the Future Emperor, Nero. He went through numerous phases in his life, becoming extremely popular, then being exiled to Corsica, then called back to Rome, he was finally accused by Nero of having taken part in a conspiracy against him even though it was extremely unlikely that Seneca was involved. Nero, his "devoted" ex- disciple, ordered Seneca to kill himself, and Seneca obeyed. Thus, following tradition, he severed several veins and bled to death.

Why is all of this important while we, as western countries, are experiencing the worst global mental suffering after the Second World War? In his short essay, "*De tranquillitate Animi* ", on the tranquillity of the psyche", a letter written to a friend who is feeling psychologically tormented, there are pieces of advice to "stay mentally healthy". Seneca gave both to his friend "Serenus" (a nickname, definitely) and to all of us readers simple and wonderful suggestions on how to stay calm, whatever the reality might be.

De tranquillitate Animi is a therapeutic book on resiliency and antifragility: in the opening, Serenus asks Seneca for counsel since he feels agitated, and in a state of unstable immobility, "as if I were on a boat that doesn't move forward and is tossed about." On one side there is a force suggesting silence and isolation, while on the other, an opposite force calling for active political engagement (while I'm writing these lines, today, in Italy, a government crisis is just opening, a wound in the flesh of this battered country). The philosopher Seneca argues that the goal of a tranquil mind can be achieved by being flexible and seeking a middle way between two extremes; a third way, not only the reiterated concept that *in medio stat virtus*, the good is in the middle*: participate to the crisis, by critically analysing the pros and cons, put the crisis into the historical frame, but don't let you so down, taking too much personally "the crisis"*.

If we want to achieve peace of mind, Seneca recommends a simple life, advises us to choose our companions carefully, since if we choose those that are corrupted by vices, their vices will extend to us. As vices, nowadays, we can think of "rage", a dominating pervasive emotion, everywhere in society, in the media, in active political engagement: if anger, for a short-limited time might be a positive and energising power for requesting, demanding and also for giving more to the world to improve the current situation, staying away from rage- poisoned people is beneficial, since this huge emotion is a destructive force both for ourselves and for others. Frugality is the main treatment for peace of mind: we have to learn to know how to contain ourselves, curb our desires, temper gluttony, mitigate anger, to look at poverty with good eyes and to revere self-control.

Seneca introduces the figure of the Stoic sage, whose peace of mind (*ataraxia*) springs directly from a greater understanding of the world. Only reasoning, caution, and foresight- vision of the future, can create the ideal atmosphere of peace in someone. The philosopher, while preserving his peace of mind, does not hate humanity for its injustice, vileness, stupidity and corruption. The times we live in are no worse than the preceding ones, and it is not reasonable to waste time raging about these evils, it is more reasonable to smile at them, and they are no worse than in the future.

The right treatment is to follow nature, find the right balance between sociability and solitude; here some of his quotes.

"Difficulties strengthen the mind, as labour does the body." –Tempering and becoming antifragile, leaving vulnerability, is a very old commandment.

"As is a tale, so is life: not how long it is, but how good it is, is what matters." The secret is adding more quality of life to years, than length of life in a depressed person. "A tale", Seneca calls in action each own personal narrative: such a modern concept, a useless longevity doesn't matter, what matters is how these years have been wisely "enjoyed". The "other" is there to help us in finding this goodness in life.

"The time will come when diligent research over long periods will bring to light things which now lie hidden. A single lifetime, even though entirely devoted to the sky, would not be enough for the investigation of so vast a subject... And so, this knowledge will be unfolded only through long successive ages. There will come a time when our descendants will be amazed that we did not know things that are so plain to them... Many discoveries are reserved for ages still to come, when memory of us will have been effaced."- and this is his legacy for the research in posterities. Human being evolves by looking at how nature works with a logical and rational mind, and what can be of help for a better life. The three main qualities of the Humanities that Seneca endorses are "Openness" to what happens, that asks for a

rational reaction, "Measure", the dosing between laughing and crying, shouting and being silent, the active and reflective mind, the expression of oneself and pure listening, and "Help" since the mere fact that he writes therapeutic books to Serenus (A friend? The emperor? And beyond, His Alter Ego?) is a sign of being a true "caring" professional.

What are the *humanities for health?* The very same ones since at least their roots are here: Seneca, already 2.000 years ago, shed light through his essays, his reflective writing, his coping attitude, in his own life endured by disease, by being a private tutor sentenced to death in the family whose destiny was tangled with murders by severing veins and poisons.

As Seneca foresaw very well, the neurosciences would later discover the positive effect of tranquillity to foster a sharp mind, to unveil reality and to move on, despite whatever reality is around: COVID-19 times, war times, normal and a new normal age, with this last being the one we long for; however, according to Seneca, we should not place such great pathos. Maybe, the new normal will come not only for ourselves in the future, but mainly for the next generations who will benefit the most.

All the dangers cannot be removed, but a safe space can be created where the body - with its neuroception - feels comfortable: the voice of its mother is what the foetus first perceives and therefore has a priority in the development of the bond between mother and child (we speak of mother tongue). To create an empathic relationship there must necessarily be a safe space generated through the voice, the facial expressions, the gaze, the being seen and considered and the contact: now, during the Covid 19 age, touch, a handshake, and the embrace are all replaced by the voice, the smile and the expression of the face.

Empathy, both emotional and cognitive, is the pillar on which to build relationships of trust that allow cooperation between individuals to overcome difficulties, build well-being, and, in a historical sense, lead to the evolution of mankind, and we need this virtue/ skill even more now to heal the traumas of the pandemic event, to become a good doctor, nurse, help professional, but more than these, good "resilient" citizens.

Tell Me Where Do the Children Play

> *"I know we've come a long way, we're changing day to day, but tell me, where do the children play?"*
> Cat Stevens

The parks were closed during the springtime of 2020. We were only allowed to walk 200 m away from our home: only one of the siblings could get out and walk for no more than 200 m, accompanies by a parent or by a relative. From March 8th to May 7th, 2020. A 5-year-old-girl met her school mate on the street: the two little girls were running towards each other to say hello, to play, but the respective parents shouted to them that their friends were dangerous.

According to the estimation of Imperial College, London, which assessed the impact of restrictions in 11 European countries - Austria, Belgium, Denmark, France, Germany, Italy, Norway, Spain, Sweden, Switzerland and the UK, the lockdown clearly saved lives - up to the beginning of May 2020. If, by that time, around 130,000 people had died from coronavirus in those countries, the researcher estimated 3.2 million people would have died by 4th May if not for measures such as closing businesses and telling people to stay at home.

That meant lockdown saved around 3.1 million lives, including 470,000 in the UK, 690,000 in France and 630,000 in Italy: these are the numbers that should be spread around, the number of lives saved [7]: that's the story to tell, not only that of the people who died. And maybe with drawings and in easy words we could change that "meme" of the 5-year-old girl being told that her friend was dangerous by reframing it, and suggesting that it was really something invisible and aggressive in the air was the real danger. But just for a period of time, a little bit longer than 40 days, perhaps. Instead of saying that the other child was unsafe.

Young children often use play, storytelling, or drawing to express their fears and wishes; this is a healthy and adaptive way for the child to try to make sense of what is happening around them. But they may also blame themselves, even if the events are out of their control, or have other inaccurate thoughts about what happened, and we have to help our child to come up with more helpful thoughts and coping statements. We must help our children identify activities to soothe themselves such as spending time with the family pet, watching a show, listening to music, or playing a favourite game.

We were hoping that this was over, but the school have reclosed and now for teenagers, undergraduate at the universities, distance learning is the daily bread and butter. We owe our youngest generations; in many cases despised since they were seen as the cause of infections and sometimes caught in the act of celebrating a "clandestine party". The young people who participated in these parties were a very tiny minority with respect to the other young students who demonstrated an extraordinary civic sentiment, for which they are sacrificing their best years behind a screen. There was a huge effort of storytelling to stress the importance of going beyond categories such *"as age or age groups, and to promote intergenerational equity and solidarity by fostering connectedness across different generations. Caring and protecting older people is not only about respecting their dignity and autonomy. It opens up invaluable possibilities for all generations to engage with collective memories and traditions – promoting the exchange of skills, knowledge and understanding"* [8]. This belief could work in the short term (in fact this article was written in May last year), but 1 year of pandemic and distance learning has led to technostress in adults, but the most endangered population in terms of mental health are the youngest generations: from a recent editorial on the British Medical Journal *"The evolving consequences of the pandemic are set against longstanding concerns about deteriorating mental health among children and young people, and the inadequacy of service provision. Although children are at lowest risk of death from covid-19, concerning signals remain about the pandemic's effects on their mental health, which are unevenly experienced across different age groups and socioeconomic circumstances"* [9].

Now parks are open again, the children are playing open air hide and seek and, on the swing, nobody dares to blame them anymore, while not enough gratitude is given to teenagers for staying apart, as lonely was considered a moral duty, and they like good soldiers have obeyed. The long Covid consequences will be seen as a post-traumatic stress disorder in their growth. I hope that not only older generations but also by the health care professionals acknowledge the younger generations because, believe it or not, if it were not for their home schooling, distance learning, online graduation ceremonies in tiny little rooms the number of elderly dead people would have been too massive to count. Praise to such responsible young people, and praise to the old people, shielding themselves at home, and waiting for so long for the vaccines have.

Here we mix together the four dimensions of the model: the biological effects, due to different social behaviour, which impacts on the psyche, which by the way can be changed by the existential values behind. Let's remember to play, as children, amidst this long uncertain time, a little every day: play for adults is critical: play has been shown to release endorphins, improve brain functionality, and stimulate creativity, and it can even help to keep us young and feeling energetic. Studies show that play improves memory and stimulates the growth of the cerebral cortex. One of the definitions of play, from the Oxford English Dictionary, is to wield lightly and freely; to keep in motion. Never rest with certainties for they are sure to grow old and vanish; and try to play by exercising the "maybe" attitude: keeping in mind two things at the same time, which are exact opposites by allowing ourselves to play ethically with thoughts, words, actions, wherever we are, whatever we face. Play is not only performing but is an action of discovery, where we can become the explorers of this new Covid world.

The Play

Twenty-two countries participated in creating a documentary "After long silence. A year remembered" [10] with texts and photos and videos centred around the experience of this Covid year with the organization created by Jonathan McFarland, "The Doctor as a Humanist." The Doctor as a Humanist is an international organization of students, scholars, educators and practitioners dedicated to the study, preservation and promotion of humanism in medicine around the world. It aims to be as international, interdisciplinary as possible but perhaps its main objective is to listen to and learn from the students and young since their initiatives, enthusiasm and energy are the driving force of this initiative whose motto is the bring back the heart and soul to medicine."

I did not want to play at the beginning. I thought I had to work more operatively, handling classes and research projects: but playing with words and images was still lingering there. And then I decided I wanted to play, and I wish to live with you in this "Play".

Spring, Summer, Autumn, Winter and Spring Again
(tribute to Kim Ki Duk,who died of Covid)

Spring 2020: or rather a month earlier, 21 February 2020: Patient 0 makes his appearance a few kilometres from Milan. He is young, 38 years old, does sport regularly and ends up in Intensive Care.

Then, it's not just about the elderly. In the following days, Northern Italy explodes, people die at home, intensive care units are not equipped, many doctors and nurses fall ill and many die from COVID. Generations of old people wiped out from towns. Yes, maybe it's more about the elderly.

The sky was beautiful, we were begging for the closure that came 2 weeks late so as not to stop production processes. But the magnolias in bloom in Milan on the blue marked the beauty of nature that didn't care about death and set its own pace.

Then they closed us down: the tiny balcony became the only place to live in contact with nature. Every flower, every seedling found or bought was taken home to carry a memory of Gea, Mother Earth, inside the house.

Summer 2020: we reopened, first inside the Region and then in Italy. Yet the squares were deserted, the people frightened, and nature in all its majesty led us into the torrid heat: wearing a mask in July on the streets was almost like not breathing in Milan, when, without a mask, breathing normally is a struggle. Only in the evenings was it refreshing. And then the drive to the SOUTH. "But has all this really happened in the North?" "YES". "The images weren't fake?" "No". The sea washed and washed, and as we drove back to the North the tears flowed knowing that the chess game with Death would begin again.

Autumn 2020: On time it was back to take more old men, and sicker young men. The trees gilded and lit up that sky which had never been so blue in Milan, even in autumn, and clean, without chemtrails; then, the first results of the vaccines. They locked us down and this time we no longer begged for it. We suffered it. We no longer knew what the point was in this dilemma: to keep the schools running? To save the lives of our elderly? Chrysanthemums were hung everywhere in Milan on the Day of the Dead, 2 November: Resist, Milano.

Winter 2020–2021: In Milan, from the La Scala theatre, arias from operas of hope and a wish for "il cielo si fa bello" "the sky is beautiful again"-"a riveder le stelle" "to see again the stars". Let's look for parks, let's breathe the air, let's walk among continuous lectures and online research. Let us not forget to get up and exercise: the discipline of the body. We eat well, we have to stay fit. Within the perimeter

Fig. 11.3 photo by Maria Giulia Marini "The magnolias in Spring 2020 and Spring 2021".

we are given. More and more restricted. From the sky, the Madonnina del Duomo looks down on us.

Spring 2021: There they are, the magnolias, 1 year later, always (Fig. 11.3) in that ironically blue sky, while 300 to 500 people die every day in Italy. We are among the last countries in terms of vaccination campaigns and Lombardy maybe will be placed under commission for ineptitude in the distribution of vaccines. They were supposed to be our lifesavers. Our reopening. A plane is taking off; I take a picture of it, and as I shoot, I feel the same thrill as when I watched the planes take off when I was a child.

I stay with my feet in the mother earth: with roots, like these trees in the park, like the bases of the columns inside the cathedrals. The two things we can visit (beyond supermarkets): parks and churches. She, Nature, goes around.

Practice Time

1. The table of the *Natural Semantic Language* is made by the following words: pick up three words by which now you are most attracted, and convert your scalene triangle to an equilateral triangle.

Category	Primes
Substantives	I, YOU, SOMEONE, PEOPLE, SOMETHING/THING, BODY
Relational substantives	KIND, PART
Determiners	THIS, THE SAME, OTHER~ELSE~ANOTHER
Quantifiers	ONE, TWO, SOME, ALL, MUCH/MANY, LITTLE/FEW
Evaluators	GOOD, BAD
Descriptors	BIG, SMALL
Mental predicates	THINK, KNOW, WANT, DON'T WANT, FEEL, SEE, HEAR
Speech	SAY, WORDS, TRUE
Actions, Events, Movement	DO, HAPPEN, MOVE
Existence, Possession	BE (SOMEWHERE), THERE IS, BE (SOMEONE/SOMETHING), (IS) MINE
Life and death	LIVE, DIE
Time	WHEN/TIME, NOW, BEFORE, AFTER, A LONG TIME, A SHORT TIME, FOR SOME TIME, MOMENT
Space	WHERE/PLACE, HERE, ABOVE, BELOW, FAR, NEAR, SIDE, INSIDE, TOUCH (CONTACT)
Logical concepts	NOT, MAYBE, CAN, BECAUSE, IF
Intensifier, augmenter	VERY, MORE
Similarity	LIKE, AS, WAY

2. Do you want to play with your pillars of the biological, psychological, social and existential model?

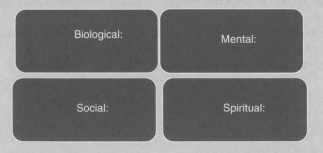

3. The application of the Natural Semantic Metalanguage to the Biological, Mental, Social, Moral Model [11]:

"Physical" (Existential) aspect:
Someone
this someone has a body
this someone can live
this someone can die this someone can do many things
many things can happen to this someone '

"Mental" aspect:
this someone can know things
this someone can want things
this someone can think about things
this someone can feel some things (feelings)
this someone can say things this someone can see things
this someone can hear things

"Social" aspect:
this someone can do things with other people
this someone can live with other people
this someone can say things to other people

"Moral" aspects:
this someone can do good things
this someone can do bad things

4. Resilience comes from the Latin word "resalio" "climb again" on a cap-sized boat… Would you like to invent another meaning for this word?
5. Quantity and quality of life? Would you like to express your idea?
6. Playing is……
7. Spring, Summer, Autumn, Winter and Spring again…. Would you like to play with circular time and narrate your experience?

References

1. Reuters Staff, Troubling "Eek" variant found in most Tokyo hospital COVID cases – NHK, April 4, 2021, https://www.reuters.com/article/us-health-coronavirus-japan-idUSKBN2BR03Wù
2. Coelho CM, Suttiwan P, Arato N, Zsido AN. On the nature of fear and anxiety triggered by COVID-19. Front Psychol. 2020;11 https://doi.org/10.3389/fpsyg.2020.581314.
3. Goddard C, Wierzbicka A. Words and meanings: lexical semantics across domains, languages, and cultures, vol. 2014. Oxford: Oxford University Press; 2016. p. 316.
4. https://www.epicentro.iss.it/coronavirus/sars-cov-2-decessi-italia.
5. Friedenberg J. Geometric regularity, symmetry and the perceived beauty of simple shapes. Empirical Stud Arts. 2017;36(1):71–89. https://doi.org/10.1177/0276237417695454.
6. Saad M, de Medeiros R, Mosini A. Are we ready for a true biopsychosocial–spiritual model? The many meanings of "spiritual". Medicines. 2017;4(4):79., Published online 2017 Oct 31. https://doi.org/10.3390/medicines4040079.

7. http://www.imperial.ac.uk/news/196234/covid-19-imperial-researchers-model-likely-impact/.
8. Carrieri D, Peccatori FA, Boniolo G. COVID-19: a plea to protect the older population. Int J Equity Health. 2020;19(1):72. https://doi.org/10.1186/s12939-020-01193-5.
9. Mental health of children and young people during pandemic. BMJ. 2021:372. https://doi.org/10.1136/bmj.n614. (Published 10 March 2021)
10. The doctors as a humanist, After long silence: a year to remember. https://youtu.be/nUZIZyd24H4.
11. Wierzbicka A. Lang Sci. 2007;29:14–65.

Glossary

Art Therapy

A therapeutic approach based on the idea that the creating process supports reparation and recovery could be defined as a form of non-verbal communication of thoughts and feelings: it can help individuals in creating meaning and achieving insight, finding relief from overpowering emotions or trauma, resolving conflicts and problems and reaching an increased sense of wellbeing. The approach is grounded on the belief that people have the capacity to express themselves creatively and that the most important thing is the therapeutic process rather than its product—so the therapist's focus is not on the aesthetic merits of art making but on the therapeutic needs of the person to express him- or herself. What is important is the person's involvement in the work, choosing and initiating helpful art activities, assisting the person in finding meaning in the creative process and facilitating the sharing of the experience of image-making with the therapist. Art therapy is acknowledged as care tool by the World Health Organization, including all visual arts, music therapy, dance therapy, creative writing, drama, story telling.

Bioethics

The term *Bioethics* (Greek *bios*, life; *ethos*, behaviour) was coined in 1926 by Fritz Jahr in an article about a "bioethical imperative" regarding the use of animals and plants in scientific research. Branch of applied ethics that studies the philosophical, social, and legal issues arising in medicine and the life sciences. It is chiefly concerned with human life and well-being, though it sometimes also treats ethical questions relating to the nonhuman biological environment. (Such questions are studied primarily in the independent fields of environmental ethics.

Bioethics emerged as a distinct field of study in the early 1960s. It was influenced not only by advances in the life sciences, particularly medicine, but also by the significant cultural and societal changes taking place at the time, primarily in the West.

M. G. Marini, J. McFarland (eds.), *Health Humanities for Quality of Care in Times of COVID -19*, New Paradigms in Healthcare,
https://doi.org/10.1007/978-3-030-93359-3

The field of bioethics has ranged from debates over the boundaries of life (e.g. abortion, euthanasia), surrogacy, the allocation of scarce health care resources (e.g. organ donation, health care rationing), to the right to refuse medical care for religious or cultural reasons.

The scope of bioethics can expand with biotechnology, including cloning, gene therapy, life extension, human genetic engineering, astroethics and life in space, and manipulation of basic biology through altered DNA, XNA (a synthetic alternative to DNA), and proteins. These developments will affect future evolution, and may require new principles that address life at its core, such as biotic ethics that values life itself at its basic biological processes and structures, and seeks their propagation.

Historian Yuval Noah Harari sees an existential threat in an arms race in artificial intelligence (AI) and bioengineering, and he expressed the need for close co-operation between nations to solve the threats by technological disruption. Harari said AI and biotechnology could destroy what it means to be human.

During Covid ages, bioethics became a major issue since the paucity of beds in Intensive Care Units, Oxygen, Vaccines had put the decision makers in the impossibility to cover all prevention and treatment care and had to decide for priorities of who to save first, and who to vaccinate first. Moreover, the Green Pass and the mandatory Covid vaccination raises futher open bioethical issues to face in community health care systems.

Coping

Coping strategies refer to the specific behavioural and psychological efforts employed to master, tolerate, reduce or minimise stressful events. Two main coping strategies have been distinguished: problem-solving strategies, that is, the proactive efforts to alleviate stressful circumstances, and emotion-focused coping strategies, which involve efforts to regulate the emotional consequences of stressful (or potentially stressful) events. Some research indicate that people use both types of strategies to combat most stressful events. The predominance of one type over another is determined, in part, by personal style and by the type of stressful event. An additional distinction, often made in coping literature, is between active and avoidant coping strategies: active strategies are behavioural or psychological responses to change the nature of the stressor itself or how one thinks about it; avoidant strategies lead people into activities or mental state that keep them from directly addressing stressful events. Generally, active strategies are thought to be better ways to deal with stressful events, whereas avoidant strategies appear to be a psychological risk factor or marker for adverse responses to stressful life events. Broad distinctions, such as problem-solving versus emotion focused, have only limited utility for understanding coping, and so research on coping and its measurement has evolved to address a variety of more specific coping strategies. Openness, responsibility, extroversion, optimism, awareness are factors fostering coping: on the contrary, introversion, ruminating thinking and denial are detrimental for coping.

COVID-19 and Long Covid

Coronavirus disease 2019 (COVID-19) is defined as illness caused by a novel coronavirus called severe acute respiratory syndrome coronavirus 2 (SARS-CoV-2; formerly called 2019-nCoV), which was first identified amid an outbreak of respiratory illness cases in Wuhan City, Hubei Province, China. It was initially reported to the WIIO on December 31, 2019. On January 30, 2020, the WHO declared the COVID-19 outbreak a global health emergency. On March 11, 2020, the WHO declared COVID-19 a global pandemic, its first such designation since declaring H1N1 influenza a pandemic in 2009. Most people infected with the COVID-19 virus experience mild to moderate respiratory illness and recover without requiring special treatment. Older people, and those with underlying medical problems like cardiovascular disease, diabetes, chronic respiratory disease, and cancer are more likely to develop serious illness, with a higher death probablity.

Long COVID is a range of symptoms that can last weeks or months after first being infected with the virus that causes COVID-19 or can appear weeks after infection. Long COVID can happen to anyone who has had COVID-19, even if the illness was mild, or they had no symptoms. People with long COVID report experiencing different combinations of the following symptoms:
- Tiredness or fatigue
- Difficulty thinking or concentrating (sometimes referred to as "brain fog")
- Headache
- Loss of smell or taste
- Dizziness on standing
- Fast-beating or pounding heart (also known as heart palpitations)
- Chest pain
- Difficulty breathing or shortness of breath
- Cough
- Joint or muscle pain
- Depression or anxiety
- Fever
- Symptoms that get worse after physical or mental activities

Ecosystem

An ecosystem is a geographic area where plants, animals, and other organisms, as well as weather and landscape, work together to form a bubble of life. Ecosystems contain biotic or living, parts, as well as abiotic factors, or non-living parts. Biotic factors include plants, animals, and other organisms. Abiotic factors include rocks, temperature, and humidity. All organisms in an ecosystem depend upon each other. If the population of one organism rises or falls, then this can affect the rest of the ecosystem, this is called interdependence.

The whole surface of Earth is a series of connected ecosystems, they are often connected in a larger biome. Biomes are large sections of land, sea, or atmosphere.

Forests, ponds, reefs, and tundra are all types of biomes, for example. They're organized very generally, based on the types of plants and animals that live in them. Within each forest, each pond, each reef, or each section of tundra, you'll find many different ecosystems.

A sustainable ecosystem is a biological environment and series of habitats that is able to thrive and support itself without outside influence or assistance. In ideal sustainable ecosystems, everything is already provided within the ecosystem for life to survive. Sustainable ecosystems across the country share several attributes, most notably biological diversity. Diversity means not only a collection of different species present, but a large amount of different species present. Other key characteristics include available acreage for roaming and expansion, available unpolluted water source and limited or controlled human activity.

There are endangered ecosystems all over the world whose diversity and sustainability is threatened daily by human actions. One of the most relevant examples of an endangered ecosystem that is quickly becoming unsustainable are coral reefs worldwide. However there are good examples of sustainable interdependence: aside from teeming forests, cities that reuse and recycle and exist in harmony with the surrounding environments are also considered sustainable ecosystems.

Humanities for Health/Health Humanities

The health humanities adopt an interdisciplinary, inclusive, applied, democratizing, and activist field of inquiry in which brings together a variety of disciplines that draws on aspects of the arts and humanities in its approach to health care, health and well-being. It involves humanities disciplines (including literature, languages, philosophy, law, politics, religion, sociology, etc) and the application of the creative arts (including music, performing arts, visual arts, film, drama, etc) to study human health and well-being.

The growth of Health Humanities throughout North America and Europe is evident in the rise of undergraduate programs, professional training opportunities, career pathways, conferences, and dedicated scholarly journals (including the British Medical Journal's Medical Humanities, Intima: A Journal of Narrative Medicine, Journal of Medical Humanities, Literature and Medicine, and Arts and Health).

Health humanities enriches students' thoughts and develops critical tools for them to critically explore artistic and cultural representations of human health, illness, and the lived effects of health policy.

Health humanities had their pioneer were with Paul Crawford in 2006, with the aim to providing a much broader application of the arts and humanities to health care, health and well-being than that currently available in fields such as medical humanities. Crawford's aim was to engage those healthcare providers, marginalized from the medical humanities, because he thought that the medical humanities have an overly narrow focus on the medicine and ruled out other disciplines related to health. Since then, health humanities moved beyond a predominating concern with training health professionals through the arts and humanities and has involved more radical and critical thinking about the concept of health and illness.

Research in the health humanities led to a variety of practical applications in terms of the health and well-being of the public, health, and social care service innovation, pedagogic developments, and policy influence.

Metaphor

Metaphor, the figure of speech that makes an implicit or hidden comparison between two things that are unrelated, but which share some common characteristics. Elena Semino, an Italian-born British linguist focused on figurative language, in Metaphor in Discourse claims that metaphor involves the phenomenon whereby we talk and, potentially, think about something in terms of something else. Over the last three decades, much attention has been paid to the presence of large numbers of metaphorical expressions in language, which we often use and understand without being conscious of their metaphoricity. In a series of influential works, the linguist George Lakoff and his colleagues pointed out that metaphorical expressions are pervasive in language and that they tend to form systematic sets such as the following: *Your claims are indefensible. He shot down all of my arguments. He's without direction in his life. I'm at a crossroads in my life.* In Metaphors We Live By (1980), Lakoff and Johnson famously showed that these expressions are not simply ways of talking about one thing in terms of another, but these set phrases are known as 'conceptual metaphors' and defined as systematic portions of our background knowledge that relate to particular experiences or phenomena (Semino 2008). Even medical language is soaked in metaphor, and this way of thinking and talking is central in medical culture and clinical practice. Many writers have looked beyond the day-to-day language to discover the basic model or the metaphor we use when thinking about medicine: *Disease is the enemy. He's a good fighter. The war on cancer. Disease is malfunction. He's in for a tune-up. Something's wrong, doc... you fix it.* It's obvious the rapid advance of the engineering and war metaphors. The literary critic and novelist Susan Sontag published a book entitled Illness and Metaphor in which she argued that the disease called 'cancer' evokes in the population a pervasive cultural myth or metaphor. Cancer is an obscene, unspeakable and shameful condition. On the other hand, medical practitioners approach cancer with a different metaphor based on military images. *Cancer is aggressive and invasive; it seeks to infiltrate and colonise by battering down the body's defences* (Sontag 1978). There are also other, more humane, metaphors for medicine, for example, physician-as-teacher. We need many such images to capture the truth, and some are more useful in healing than others; indeed, everyday medicine is replete with evidence of the power of language and narrative to heal or to harm (Coulehan 2003). Metaphors are a useful device to grasp an unknown concept by using a known concept, thus naming and explaining a phenomenon, i.e. illness, which otherwise would remain unintelligible, obscure. They make it easier to have a broader understanding of the disease in its aspects not only biological but also psychological and emotional, helping to grasp meanings that are otherwise not expressed and not said.

Narrative Medicine

There are many different definitions of narrative medicine given by experts. According to what was written by Brian Hurwitz and Trisha Greenhalgh, it's everything that happens between the healthcare profession and the patient. According to Rita Charon, it enters in the depth of daily clinical practice, and it is based on the skill of the professionals to observe, listen, interpret and being moved by the patient's narrations. According to John Launer, narrative medicine means understand if there are changes after the beginning of the disease, or if the stories remain 'stopped', with the patient's trauma facing the new undesired condition. The Superior Institute of Healthcare with the CNMR (National Centre for Rare Diseases) defines that with the term narrative medicine, a methodology of clinical-helpful intervention based on the specific communicative skill. According to Maria Giulia Marini narrative medicine it is based on authentic narratives, patients', carers' citizens' experiences. It is a care approach more focused on the whole human being, than just on the bodies abnormalities given by the disease: the analysis of the text written by patients, by healthcare professionals and by relatives allows to understand the culture, the values, the needs, the passions, the personal and the professional projects, and it's on this that we concentrate to create or maintain an healthcare ecosystem. Narrative medicine rewrites the medical and scientific terminology to make it more coherent with the living and the way of thinking of the patient. Conceptually, narrative medicine goes beyond the 'artistic representation of the stories of illness' and becomes a science that helps all the healthcare system professionals and policy makers, giving attention the experiences of people who live with a pathology and of their caregivers, through the research and the clinical practice. It is able to join ill people and healthcare operators, to associate Evidence-Based Medicine and Narrative-Based Medicine, as well as clinical sciences and human sciences. Narrative Medicine belongs to anyone involved, both as patient and as carer, in the therapeutic process, to create more wellbeing and a safe space even in the most troublesome situations.

Natural Semantic Metalanguage (NSM)

It is based on a shared core of simple words called semantic primes that could be translated in many different languages evoking exactly the same concept. It was the main result of an intensive studies lasting beyond 30 years by Anna Wierzbicka,

Professor of linguistics at the Australian National University in Canberra University, Australia. Sixty-five semantic primes emerged from her long-lasting study, a group of words that could be used as linguistic and cultural analysis tool to explain the meaning of more complex sentences or cultural-specific concepts as values and cultural attitudes. Actually, the NSM represents a universal language that allows us to formulate analysis: firstly, reducing the risk of any sort of misunderstanding and secondly, without the problem of traducing from the 'Anglo-centric' sense of meaning.

The 65 primes words of the Natural Semantic Metalanguage.

I, me, you, someone, something-thing, people, body	Substantives
Kind, part	Relational substantives
This, the same, other, else	Determiners
One, two, much, many, little, few, some, all	Quantifiers
Good, bad	Evaluators
Big, small	Descriptors
Think, know, want, don't want, feel, see, hear	Mental predicates
Say, words, true	Speech
Do, happen, move, touch	Actions, events, movement, contact
Be (somewhere/someone/something), there is, is mine	Location, existence, specification, possession
Live, die	Life and death
When, time, now, before, after, a long/short time, for some time, moment	Time
Where, place, here, above, below, far, near, side, inside	Space
Not, maybe, can, because, if	Logical concepts
Very, more	Augmenter, intensifier
Like	Similarity

Pandemic

The root of the word Pandemic is (from Greek πᾶν, pan means "all" and δῆμος, demos means "local people" the "crowd') so it is an epidemic of an infectious disease that occurring on a scale that crosses international boundaries, usually influencing people on a worldwide scale. Based on oxford advanced dictionary, pandemic is a disease that spreads over a whole country or the whole world.

Based on Merriam Webster, pandemic as an adjective means: "occurring over a wide geographic area (such as multiple countries or continents) and typically affecting a significant proportion of the population" and pandemic as a noun means: "an outbreak of a disease that occurs over a wide geographic area (such as multiple countries or continents) and typically affects a significant proportion of the population: a pandemic outbreak of a disease".

Although the definition of pandemic and epidemic seems the same, there are also some differences. An epidemic is an outbreak of disease that spreads quickly and affects many individuals at the same time. A pandemic is a kind of epidemic: one which has spread across a wider geographic range than an epidemic, and which has affected a significant portion of the population.

A disease or condition is not a pandemic merely because it is widespread or kills many people; it must also be infectious. For instance, cancer is responsible for many deaths but is not considered a pandemic because the disease is not contagious (i.e. easily transmittable) and not even simply infectious.

Throughout human history, there have been a number of pandemics of diseases. The most terrible and fatal pandemic in recorded history was the Black Death in the fourteenth century (also known as The Black Death), which killed an estimated 75–200 million people. The 1918 influenza pandemic (Spanish flu) is another example of pandemics. Current pandemics include COVID-19 (SARS-CoV-2) and HIV/AIDS.

Pandemic Fatigue

Pandemic fatigue is the state of being worn out by recommended precautions and restrictions relating to a pandemic, often due to the length of the restrictions and lack of activities for one to engage in, resulting in boredom, depression, and other issues, thereby leading one to abandoning these precautions and risk catching the disease.

In another definition: a natural and expected reaction to sustained and unresolved adversity in people's lives. Expresses itself as emerging demotivation to engage in protection behaviours and seek COVID-19-related information and as complacency, alienation and hopelessness. Pandemic fatigue results from various barriers that people experience across cultural and national contexts, and that require different kinds of support, structures and communication.

Signs of pandemic fatigue:

1. Eating or sleeping more or less than usual
2. Having trouble focusing
3. Feeling edgy or nervous
4. Snapping at or arguing with others
5. Lack of motivation
6. An inability to stop racing thoughts
7. Withdrawal from others

There are some harm-reduction approaches that recognize the harms of pandemic fatigue.

Telemedicine

Telemedicine is a term coined in the 1970s, which literally means "healing at a distance". Telemedicine refers to the provision of remote clinical services, via real-time two-way communication between the patient and the healthcare provider, using electronic audio and visual means. Physicians and patients can share information in real time from one computer screen to another. And they can even see and capture readings from medical devices at a faraway location. Using telemedicine software, patients can see a doctor for diagnosis and treatment without having to wait for an appointment. Patients can consult a physician at the comfort of their home. There are three common types of telemedicine, which include but not limited to:

1. Interactive Medicine—which allows patients and physicians to communicate in real-time while maintaining HIPAA (The Health Insurance Portability and Accountability Act) compliance;
2. Store and Forward—which permits providers to share patient information with a practitioner in another location it does not require the presence of both parties at the same time. Dermatology, radiology, and pathology are common specialties that are conducive to asynchronous telemedicine.
3. Remote Patient Monitoring, also known as self-monitoring or testing, – which allows remote caregivers to monitor patients that reside at home by using mobile medical devices to collect data (e.g. blood sugar or blood pressure).

Telemedicine is one of the main domains of a wider scope named "Digital health". Digital health is the combination of digital technologies with health and healthcare to improve the efficacy of healthcare delivery. Digital health includes categories such as mobile health (mHealth), health information technology (IT), wearable devices, telehealth and telemedicine, and personalized medicine.

Digital health technologies include both hardware and software solutions and services, from mobile medical apps and software to artificial intelligence and machine learning which give a more holistic view of patient health through access to data and giving patients more control over their health.

Wellbeing

Wellbeing is a growing area of research, but the question of how it should be defined remains unanswered: there is a greater necessity to be clear about what is being measured and how data should be interpreted, in order to undertake a valid assessment. Any new definition must go beyond an account or description of wellbeing itself and be able to make a clear statement of the meaning of the term. The knowledge of the historical background of the study of wellbeing can be necessary for this purpose, and one can focus on two main approaches—the hedonic tradition, which accentuated constructs such as happiness, positive affect, low negative affect and satisfaction with life, and the eudaimonic tradition, which underlined positive psychological functioning and human development. Despite these two different approaches, most researchers now believe that wellbeing is a multidimensional construct—and, consequently, the diversity dimensions have created a confusing and contradictory research base. To move closer to a new definition of wellbeing, we may focus on three key areas: the idea of a set point for wellbeing, the inevitability of equilibrium/homeostasis and the fluctuating state between challenges and resources. Consequently, the research has proposed a new definition of wellbeing as the balance point between an individual's resource (psychological, social and physical) pool and the challenges (psychological, social and physical) faced.

Gratitude and thanks to Saba Mirikermanshahi and Negin Nouraei for their great work in editing and putting the Glossary together.

Biographies of the Editors and Authors

Maria Giulia Marini, Epidemiologist, counsellor in transactional analysis. More than 30 years of professional life in health care.

Classic humanistic background, followed by scientific academic studies, chemistry and pharmacology, in Milan. Worked in clinical research, moved to health care organization, getting academic specialization in Epidemiology. Currently, director of Innovation in Health Care Area of Fondazione ISTUD, with a humanistic approach acknowledged by the Italian Ministry of Research. Founder and Serving President in 2020 of EUNAMES, European Narrative Medicine Society and tenured professor of Narrative Medicine at La Sapienza, Rome, and in 2016, referee for World Health Organization for "Narrative Method in Public Health."

Writer of the book in 2016; "Narrative medicine: Bridging the gap between Evidence Based care and Medical Humanities," with Springer. Last published book in 2019, with Springer "The Languages of care in narrative medicine: words, space and sounds in the healthcare ecosystem". In the same year, diploma in Art-coach. Editor in chief of the on-line journal: *Chronicles of health care and narrative medicine.* More than 100 publications on narrative medicine on peer reviewed journals and local press and author on the most prestigious Italian Encyclopaedia Treccani on "Narrative Medicine" and Empathy" in 2020. Lecturer in different international contexts from Academies to Public and Private Foundations.

Motto: We are such stuff as dreams are made on.

M. G. Marini, J. McFarland (eds.), *Health Humanities for Quality of Care in Times of COVID -19*, New Paradigms in Healthcare, https://doi.org/10.1007/978-3-030-93359-3

Jonathan McFarland, Trained in Linguistics and Literature, Jonathan McFarland currently works as Associate Professor in the Faculty of Medicine at Universidad Autonoma de Madrid and as a Senior Lecturer at I.M. Sechenov First Moscow State Medical University.

He is President of the newly formed The Doctor as Humanist association, which is an international organization of students, scholars, educators and practitioners dedicated to the study, preservation and promotion of humanism around the world. The association runs a yearly online course, which is entering its 4th edition as well as holding face-to-face (2017, Mallorca; 2019, Moscow) and virtual conferences.

As well as presiding over The Doctor as a Humanist, he is also on the board of EUNAMES (European Narrative Medicine Society), a member of COMETA (Ethics Committee for Primary Care in Mallorca), and an honorary member of the Gimbernat Surgical Society.

His interests are varied but include Medical Writing, Medical English, Medical Humanities and Medical Education in general, and within the Medical Humanities he has a particular interest in literature and medicine, narrative medicine as well as the philosophy of medicine and how the arts, in general, can help medical students and young doctors learn to look for the person in the patient.

Motto: Let us learn to embrace the uncertainty around us

Biographies of the Authors

David Cerdio, Medical Doctor graduated from Universidad Anáhuac México, Specialist in Health Management and Corporate Welfare, Masters in Management and Administration of Health Institutions, Algology Master student and Applied Bioethics PhD Candidate. Fervent promoter of medical humanism and humanities as a way to acquire such competence. Basic Science Coordinator at Universidad Anáhuac México, Medical Director at Luz en la Calle humanitarian aid Institution, General Secretary at The Doctor as a Humanist International Association.

Motto: Transforming the world is just matter of starting somewhere.

Paola Chesi, University degree in Natural Science, with expertise in marine biology. Researcher and educator at Healthcare and Wellbeing Area of Fondazione ISTUD since 2010. Previous experience in environmental associations in the University of Turin. Interests in narrative medicine applied to healthcare, healthcare providers' wellbeing, diversity inclusion, analysis of the language and story-telling.

Coordinator of educational programs and Master addressed to healthcare professionals, and future expert in medical humanities with lectures in narrative medicine. Member of the Board of European Narrative Medicine Society - EUNAMES. Author of dozens of scientific and educational publications on narrative medicine on peer reviewed journals.

Motto: a falling tree makes more noise than a growing forest.

M. G. Marini, J. McFarland (eds.), *Health Humanities for Quality of Care in Times of COVID -19*, New Paradigms in Healthcare, https://doi.org/10.1007/978-3-030-93359-3

Carol Ann Farkas, PhD, is Professor of English in the School of Arts and Sciences at MCPHS University, where she directs and teaches in the first-year English program, and teaches elective and directed study courses in nineteenth-century fiction and narrative and medicine. As a scholar with a background in Victorian, cultural, and composition studies, she currently specializes in the interdisciplinary study of medicine and wellness in popular culture. Previous publications have focused on the gendered representation of physical competence in popular fitness magazines ("Bodies at Rest, Bodies in Motion: Women's Fitness and Physical Competence" in Genders Online); the wellness "education" constructed by fitness magazines ("Tons of Useful Stuff": Defining Wellness in Popular Magazines," in (Studies in Popular Culture); the teaching of literature and medicine in a health sciences context ("Teaching Madness and Literature in a Healthcare Context: An Enquiry into Interdisciplinary Education" in Mental Health Review Journal, co-authored with David Flood); the ways in which popular media contribute to the experience of psychosomatic illness ("Potentially Harmful Side-Effects: Medically Unexplained Symptoms, Somatization, and the Failed Illness Narrative for Viewers of Mystery Diagnosis," in The Journal of Medical Humanities, and "The Blind Men and the Elephant: Mediating Conflicting Views of Controversial Diagnoses" in Trespassing.); and the treatment of intelligence in popular culture ("What's the Difference?: Pathologizing Genius and Neurodiversity in Popular Tele-vision Series"). She is a past president of the Northeast Popular Culture Studies Association and is currently the chair of the Health, Disease, and Wellness Area for the Popular Culture Association in the US. She is also a member of the Centre for Health Humanities at MCPHS University, a group which fosters the interdisciplinary scholarship and teaching of the medical humanities. In 2020 she became the director of the Centre's new undergraduate degree program in health humanities. Her most recent publication is an edited anthology published in 2017 by Routledge: Reading the Psychosomatic in Medical and Popular Culture: Something, Nothing, Everything. Forthcoming work includes book chapters on eco-anxiety in popular media, and a study on the role of reading during the pandemic.

Motto: Hint: the cage is not locked (Nove Knutson).

Fabrizio Gervasoni is a physician and physiatrist at the Physical Medicine and Rehabilitation Unit, a department of the "Luigi Sacco" University Hospital, A.S.S.T Fatebenefratelli-Sacco, Milan, Italy. During the past 18 months he has been treating many patients requiring Post-COVID rehabilitation care.

Fabrizio is also a journalist. Since October 2018, he is the Editor in Chief of the medical journal published by Springer Healthcare "Medici Oggi", a periodical founded in 1997 and very well-known amongst Italian physicians. He is also a board member of the Doctors Guild of Milan. Apart from his professional activities, Fabrizio plays keyboards in a band called the "Milanestrone", a fantasy term merging the words "Milan" and "Minestrone", a popular soup in the region.

Motto: No matter which tools, methodologies and points of care are considered, an empathic relationship with patients will always represent the most important instrument for a clinician in order to define the most appropriate diagnostic and therapeutic pathways.

Stephen Legari holds a master's degree in Creative Arts Therapies from Concordia University and a master's degree in Couple and Family Therapy from McGill University and is a registered art therapist, licensed psychotherapist and family therapist. Since 2017, Stephen has worked at the Montreal Museum of Fine Arts in Montreal, Canada as Program Officer for Art Therapy in the Division of Education and Wellness. This program specializes in community and clinical partnerships that seeks to provide arts-based therapeutic activities for a range of publics within a fine art museum context. Stephen has also supervised masters level students in art therapy and social work and has published several texts on museum-based wellness and therapy practice within a broader movement of the arts in health.

Motto: Kindness is radical.

Susana Magalhaes Team Coordinator of the Unit for Responsible Research at Institute for Research and Innovation in Health (Instituto de Investigação e Inovação em Saúde) (i3S), University of Porto; PhD holder in Bioethics (Portuguese Catholic University) with a dissertation entitled "Bioética e Literatura: entre a Imaginação e a Responsabilidade" (If-Press, 2016); and a researcher in the areas of Narrative Medicine and Research Ethics/Integrity. Lecturer at University Fernando Pessoa and invited lecturer at Portuguese Catholic University.

Her previous research activity at the Institute of Bioethics—Portuguese Catholic University (IB-UCP) included writing scripts for a series of documentaries on Science, Ethics and Society (https://www.dgs.pt/em-destaque/documentarios-bioetica-como-estrategia-para-a-participacao-do-publico-em-materias-cientificas.aspx).

She is author of several articles on Bioethics and Narrative Medicine and she has been coordinating a Group of Study and Reflection on Narrative Medicine (Grupo de Estudos e Reflexão em Medicina Narrativa—GERMEN) since 2018.

Motto: Listen with others' eyes and see with other' stories: the beauty of the world lies within.

Albina Vegel is a researcher, an educator, and a project coordinator with 7 years of working experiences in higher education management institutions. She has lived and worked in Slovenia, USA, Spain, and the Netherlands where she currently resides and works. Ms. Vegel speaks English, Spanish, Slovene and Dutch. She obtained her BA in journalism at the University of Ljubljana in Slovenia. In 2020 she graduated from MA in Management of Organizations in the Knowledge Economy from Universitat Oberta de Catalunya in Spain with a thesis titled Sustainability Consciousness among Students of Pompeu Fabra University for which she was awarded a maximum grade with honours. She also presented her work at an online symposium New Realities in Times of Covid-19 organized by The Doctor as Humanist in November 2019. She aspires to continue working with students and professors in the education sector as a project manager and at the same time to raise awareness about sustainability and sustainable development matters among the academic and wider community. Now, happily working at Rotterdam University.

Motto: Per aspera ad astra.

Neil Vickers Professor of English Literature and the Health Humanities at King's College London. He is also the codirector of the Centre for the Humanities and Health there. He worked for many years in epidemiology and public health, and has a strong interest in all aspects of the psychic aura of physical illness. His main interest is in literature and medicine. He has published widely on the application of psychoanalytic psychosomatics to first-person memoirs of illness, and on illness narrative as a new genre of life-writing. He is the author of Coleridge and the Doctors (2004). He is currently writing a book (with Derek Bolton of the Institute of Psychiatry, Psychology and Neuroscience) called Shared Life and the Experience of Illness.

Motto: 'Give every child the best start in life. Enable all children, young people and adults to maximise their capabilities and have control over their lives. Create fair employment and good work for all. Ensure a healthy standard of living for all. Create and develop healthy and sustainable places and communities' (Michael Marmot).

Printed in the United States
by Baker & Taylor Publisher Services